THE COMPETITIVE CHALLENGE

Strategies for Industrial Innovation and Renewal

**BALLINGER SERIES
ON
INNOVATION AND ORGANIZATIONAL CHANGE**

Series Editors:

Michael Tushman and Andrew Van de Ven

THE COMPETITIVE CHALLENGE

Strategies for Industrial Innovation and Renewal

Edited by DAVID J. TEECE
The Transamerica Lectures in Corporate Strategy
School of Business
University of California
Berkeley

BALLINGER PUBLISHING COMPANY
Cambridge, Massachusetts
A Subsidiary of Harper & Row, Publishers, Inc.

International Standard Book Number: 0–88730–178–9

Library of Congress Catalog Card Number: 87–1190

Printed in the United States of America

Library of Congress Cataloging-in-Publication Data

The Competitive challenge.

   "The Transamerica lectures in corporate strategy,
school of Business, University of California, Berkeley."
   Includes index.
   1. Industrial management—United States.   2. Strategic
planning—United States.   3. Technological innovations—
United States—Management.   4. Competition, International.
I. Teece, David J.   II. Transamerica Corporation.
HD70.U5C58   1987   658.4'06   87–1190
ISBN 0–88730–178–9

CONTENTS

LIST OF FIGURES

LIST OF TABLES

PREFACE

This book is a collection of lectures presented at the School of Business, University of California, Berkeley, under the sponsorship of the Transamerica chair in corporate strategy. The lectures were initially presented during the academic year 1984–85 but in most cases were subsequently revised and extended. The generous support of the Transamerica Corporation in endowing the chair and permitting the funds to be used to support the lectures is gratefully acknowledged. Angela Jansa's careful attention to the organizational detail associated with the lecture series is also deeply appreciated. The active support and participation of faculty, in particular, of David Aaker, Fred Balderston, Robert Harris, Dick Holton, and Charles O'Reilly, also helped make the program a success. Dean Raymond Miles's support and encouragement were instrumental to the launching of this endeavor.

The lecture series, "Strategy and Organization for Industrial Innovation and Renewal," brought together leading scholars of strategy and organization and required them to address issues in international competition, either at an applied or at a theoretical level. This mix of theory and application is deliberate. The strategy field,—and to a large degree the organization field as well,—is immature; and robust empirically valid theoretical frameworks that can be applied to normative issues are lacking. Hence, it made sense to permit exploration into fundamental issues in strategy, organization, and entrepreneurship, particularly by scholars who had already demonstrated their abilities to make major contributions to their fields.

The first part of this book examines U.S. industry in international competition, and the second part contains largely theoretical expolorations of issues central to the competitiveness of the business enterprise. Although the primary focus of several of the chapters is on the U.S. experience, most of the material, even in Part I, is relevant to competitive strategy for any firm exposed to international competition.

1 INTRODUCTION

David J. Teece

The environment in which business operates has changed dramatically over the past two decades, particularly for U.S. business. The U.S. economy is much more interdependent than it used to be. Imports and exports now represent twice as large a portion of U.S. GNP as they did two decades ago. Almost one fifth of U.S. industrial production is exported, and fully 70 percent of the goods produced in the United States compete with merchandise from abroad (President's Commission on Industrial Competitiveness 1985).

New forms of competition and new competitors now challenge the commercial position of U.S. firms at home and abroad. These new competitors (especially Japan) are throwing the lackluster performance of U.S. firms, particularly those that manufacture, into sharper relief. Table 1–1 summarizes productivity growth rates in manufacturing for seven advanced industrial economies, and U.S. performance is even worse than that of the United Kingdom. Management practices and policies that may have sufficed in the past are no longer providing or supporting the competitive advantage that has been the foundation of a superior standard of living in the United States.

This book attempts to come to grips with several dimensions of the competitive challenge. In the main, the authors collectively recognize that strategic choices—decisions that have implications of major importance for the success and survival of the business firm—can benefit from an understanding of international markets, organizational behavior, skill accumulation, and the innovation process, particularly as it relates to manufacturing. They also recognize that traditional country-centered

Table 1–1. Growth Rates of Manufacturing Output per Hour

	1950–57	1957–66	1966–73	1973–79	1979–83	1950–83
Japan	9.50	8.17	11.25	6.35	6.98	8.63
West Germany	6.99	6.37	5.43	4.40	2.46	5.47
Italy	5.82	6.51	6.82	3.25	3.04	5.42
France	4.35	6.12	6.26	4.99	3.87	5.29
Canada	3.59	4.19	4.88	2.20	1.13	3.48
United Kingdom	1.42	3.42	4.95	1.52	3.54	2.99
United States	2.07	3.07	3.02	1.46	2.71	2.51

Source: U.S. Bureau of Labor Statistics (1985).

economic concepts such as comparative advantage do not assist firms as they seek achievement of world market leadership. Although a national enterprise may start with a comparative advantage-related edge derived from national endowments, this edge must be translated into a broader array of advantages that arise from a global approach to configuration and coordination. Firms can build distinctive international competences that transcend their national resource base. In short, strategy does matter, although in certain instances substitutes can be crafted (Chapter 10).

Organization also matters mightily. The range of assets that a firm controls (scope) and the form and character of its organization (structure, culture) also affect performance. Wide variations in the productivity of firms in the same industry using the same technology and serving similar customers are evidence of the importance of human resources and their effective management. Strategy and organization are key determinants of performance and are at least to some degree subject to managerial influence.

However, superior planning and strategizing are necessary but not sufficient conditions for superior performance. The wrong organizational structures, incentives, and cultures can completely confound strategy. The right structures and cultures can cause strategies to be implemented more quickly and cheaply. Inferior factors of production can in turn confound the best strategies and organizational structures. Although many of these important issues are not dealt with in these chapters, common themes do appear.

It should also be clear that this volume is primarily about corporate strategy, not national strategy. The latter has come to signify the set of public policies that national governments sometimes fashion in order to pursue certain industrial goals. For example, Japanese firms, encouraged by government, have apparently targeted supercomputers and biotechnology as strategic industries. This volume is primarily concerned with strategy

at the firm level, although several chapters recognize the importance of consistency between public policy and corporate strategy. Indeed, one chapter points to the desirability of viewing public policy and strategy together.

A central theme of this volume is the importance of technological innovation and entrepreneurship to U.S. performance. More so than in any other economy, the United States today depends critically on its ability to innovate—and to capture the benefits from innovation—for its economic prosperity. Yet that key capability is seriously threatened by weaknesses in the economic infrastructure, by a failure to lead in process technology, and, most important, by fiscal and monetary policies that raise the cost of capital and crimp the innovator's ability to capture return from innovation.

Although the U.S. predicament—one that is mirrored in many other advanced industrial countries—forms the backdrop for the inquiry initiated in this volume, the second half of this book explores issues in strategy and organization at a fundamental level. One chapter points to a current deficiency in strategic thinking—the inability to incorporate systematically the external environment and interfirm relations in strategic analysis. Two others explore strategy and entrepreneurship at an abstract level and seek to provide the underpinnings of a theory that can further the systematic understanding of the development and growth of the business enterprise. At a time when so much attention is given to innovation and entrepreneurship, it is rather pathetic that a deep understanding of the process is lacking. It is no wonder that firms and governments have difficulty trying to stimulate entrepreneurship and innovation when the factors that propel it are so poorly understood.

A thumbnail summary of each chapter follows, coupled with minor editorializing. Readers are urged to explore the individual chapters to capture their full richness and diversity. These summaries represent best efforts to understand the chapters, and the interpretations and embellishments herein ought not to be attributed to the authors.

Porter (Chapter 2)

In "Changing Patterns of International Competition," Michael E. Porter explores the implications of increased international competition for competitive strategy. He identifies several distinctive issues caused by international competition that are not raised by domestic competition.

Using the industry as the unit of analysis, Porter distinguishes between industries that he labels *multidomestic* and those that he labels *global*. The former consists of a diminishing breed of industries, largely services (such

as dry cleaning, electric power, local telecommunication) where competition in each country or region is largely independent of competition in other countries. In this type of industry, multinational firms may exist, but they modify and adapt their capabilities to each market. Outcomes are largely determined by conditions in each country.

In global industries, on the other hand, a firm's competitive condition in one country is significantly affected by its position in other countries. Therefore, an international industry is not merely a collection of domestic industries but a series of linked domestic industries in which rivals compete against each other on a truly worldwide basis. Examples include commercial aircraft, television sets, copiers, watches, and personal computers.

In multidomestic industries, strategy may be country centered, while in global industries a firm must integrate activities on a worldwide basis to capture linkages and economies. Managing international activities in a highly decentralized fashion like a portfolio, as is possible in a multidomestic industry, will impair competitive performance. That is not to say that local factors can be overlooked in global industries; rather, a careful balancing of local autonomy with global strategy is required.

In order to ascertain whether there is an advantage to integrating activities on a worldwide scale requires, for purposes of analysis, the horizontal disaggregation of the firm activities into its discrete parts. An appropriate tool for this is Porter's *value chain*, which recognizes nine categories: firm infrastructure, human resource management, technology development, procurement, inbound logistics, operations, outbound logistics, marketing and sales, and service. These activities are connected through *linkages*—the way one activity is performed frequently affects the cost or effectiveness of other activities. The firm's value chain is embedded within a larger stream of activities termed the value system; the connections here (such as with suppliers) also become essential to competitive advantage.

A firm that competes internationally must decide how to spread the activities in the value chain geographically. Downstream activities are usually tied to the buyer and hence create competitive advantages that are largely country specific: A firm's reputation, brand name, and service network in a country grow out of a firm's activities in that country and create country-specific entry/mobility barriers. Competitive advantage in upstream and support activities, on the other hand, grows more out of an entire system of countries on which a firm competes than out of its position in any one country. The location and scale of these footloose activities must be optimized on a worldwide basis to achieve competitive advantage. The international operations must then be integrated effectively.

When firms seek to gain competitive advantage through astute international configuration and superior coordination, they are pursuing, in

Porter's terms, a global strategy. There are structural aspects of an industry that work for and against globalization. Economies of scale, proprietary learning, and national comparative advantage favor locating production in a single country. Local product needs, transportation costs, and government policy work the other way and favor dispersion. Dispersion, in turn, imposes organizational challenges because country subsidiaries must operate in collaboration, not in competition. Advantages from configuring operations to local conditions can be lost, however, without the appropriate organizational structures and mechanisms to tie far-flung operating units together.

Commercial aircraft represents an extreme case of a global industry because there are three competitors, each with a global strategy: Boeing, McDonnell Douglas, and Airbus; but even here, globalization is tempered by offsets. Alumina is globalized upstream (smelting) but country-centered downstream (fabrication). In global industries first movers often lead because they gain scale and learning advantages that are difficult to overcome. Global leaders often begin with some advantage at home, although the global strategy creates new advantages that are often more durable.

When coupled with the value chain apparatus, the globalization versus multicountry strategy can lead to complex strategies involving a mixture of home-country and global approaches. Whereas the traditional approach often involves characterizing strategy as a choice between worldwide production centralization and local tailoring, the Porter framework involves a search for competitive advantage from configuration/coordination at each stage of the value chain. A firm may concentrate some activities while dispersing others.

When the pattern of international competition shifts from multidomestic to global, there are important implications for the strategy of international firms. In global industries, concerns relating to the firm's need to adapt to local conditions and ways of doing business must be supplemented with an overriding focus on the ways and means of international configuration and coordination. In government relations, for instance, the focus must shift from stand-alone negotiations with host countries to a recognition that negotiations in one country will both affect other countries and be shaped by possibilities for performing activities in other countries. In finance, measuring the performance of subsidiaries must be modified to reflect the contribution of one subsidiary to another's cost position or differentiation in a global strategy, instead of viewing each as a stand-alone unit. In battling with global competitors, it may be appropriate in some countries to accept low profits indefinitely—in multidomestic competition this would be unjustified. In global industries, the overall system matters as much or more than the country.

Cooperative agreements (coalitions, joint ventures, partnerships, and so forth) take on special importance within a globalization strategy because they are a way of configuring activities in the value chain on a worldwide basis jointly with a partner. Coalitions are a natural consequence of globalization and the need for an integrated worldwide strategy. The same forces that lead to globalization prompt the formation of coalitions as firms confront barriers (such as access to foreign markets, scale and learning thresholds) to establishing a global strategy on their own. In many industries, coalitions can be a transitional state in the adjustment of firms to globalization, reflecting the need of firms to catch up in technology, cure short-term problems caused by exchange rates, and accelerate foreign-market entry. These relations are often exposed to hazards that managers are slow to recognize.

Japan has been the clear winner in the postwar globalization of competition. Japanese multinationals had the advantage of embarking on international strategies in the 1950s and 1960s when the imperatives for a global approach to strategy were beginning to accelerate, but without the legacy of past international investments and modes of behavior. Most European and many U.S. multinationals, on the other hand, were well established internationally before the war and accordingly possessed legacies of local subsidiary autonomy that made it difficult to coordinate globally. The U.S. challenge is to catch up to the Japanese in a variety of technologies, as well as to learn how to gain the benefits of coordinating among dispersed units instead of becoming trapped by the myths of decentralization.

Clark (Chapter 3)

Clark points out that changes underway in technology and international competition will enhance the importance of innovation in both established and emerging manufacturing industries and redefine its competitive role. New concepts in products and processes have altered the value of existing commitments and created new possibilities in marketing and production. The result has been intense competitive pressure on established firms using old technologies and competing according to a different paradigm.

Fortunately, the forces driving change create opportunities as well as challenges. Success can flow from a commitment to new concepts in managing production, new approaches to customers and markets, and the application of new technology. These changes require investment of a special kind.

An important new concept introduced by Clark is that of transilience. The foundation of a firm's competitive position is the set of capabilities the firm uses to build the product features that appeal to the market. The significance of a change in technology depends on its capacity to influence the firm's existing resources, skills, and knowledge—its *transilience*. An innovation has low transilience if it enhances the value or applicability of the firm's existing capability, and high transilience if it disrupts and destroys.

The impact of technology can be measured according to its effect on both the market and the technology: (1) Architectural innovation occurs when radical technology is applied to new markets (such as the Bessemer process in steel production);[1] (2) niche-creating innovation (such as continuous heat treating)[2] occurs when refinements in technology are applied to new customer groups and new applications; (3) regular (incremental) innovation occurs when refinements in technology are applied to existing markets and customers (such as modern continuous casting);[3] (4) revolutionary innovation (such as the continuous micromill)[4] follows from a disruptive change to technology but not to markets or customers. Clark points out that these different kinds of innovations require different competitive strategies and depend on different kinds of managerial and organizational support. Moreover, the four types of innovation differ in the way they create both value and uncertainty and in the supporting investments they require.

The transilience framework has powerful implications for organization and management and particularly for the use of project evaluation tools. For instance, a mechanical application of discounted cash flow (DCF) analysis will be helpful for analyzing a regular innovation but may be misleading for niche-creating innovations (it may obscure new market opportunities). Furthermore, the transilience framework can be useful to augment traditional competitive analysis by providing a framework for assessing the relative impact of a new technology on industry participants. Except perhaps in the case of regular innovation, the framework also stresses the importance of a multidisciplinary general management perspective to decisionmaking with respect to innovation.

Wheelwright (Chapter 4)

Wheelwright forcefully argues that rising labor costs are not the primary cause of declining U.S. manufacturing competitiveness. Rather, the most important explanation for the worldwide decline in U.S. manufacturing competitiveness is management's view of the manufacturing function, its role, and how that ought to be carried out. Restoring the competitive edge requires a basic change in philosophy, perspective, and approach. Widely held U.S. views of manufacturing are, quite simply, incorrect.

The traditional management view of manufacturing Wheelwright characterizes as *static optimization*. This form of management is strongly hierarchical, leaving little to worker initiative. It has its intellectual roots in Taylorism, in which management determined the "best" way for a worker to perform a task. This system requires the commitment of large resources to monitoring and coordination. It also leads to rigidities because a top-down approach to management works best when the environment is stable and predictable. Wheelwright associates some important disabilities with this approach, not least of which is a decline in worker motivation and initiative, disenchantment, and alienation.

Perhaps as a consequence of a desire for predictability, the traditional system leads to an artificial and costly dichotomy between product development and manufacturing process development. Thus, new products are developed in sequential fashion with R&D taking full responsibility initially and then tossing that responsibility "over the wall" to manufacturing at a later stage. Another unfortunate consequence of this approach is that new manufacturing technologies are considered of secondary importance—an afterthought lacking comparable competitive significance to product development. Process technologies get developed not in anticipation of new product opportunities but rather only as required by the pricing realities of the marketplace (costs must be lowered) or the product characteristics as defined by R&D (the existing processes cannot be used to make the new product). Manufacturing expertise thus comes to reside outside the organization—in the equipment suppliers—where it cannot be a source of the firm's or the nation's competitive advantage.

A further consequence is that cost overruns and delays too often accompany the commercialization of new technology. Because designs passed from R&D to manufacturing are rarely informed by manufacturability concerns, new products often have to be reengineered, compromising the timely and cost-effective introduction of new technologies. Current manufacturing practice thus too often keeps manufacturing in a responsive mode, with little opportunity to carry out long-term systematic developments. In this context, it is difficult for manufacturing to become a source of competitive advantage.

Wheelwright advocates a different strategic vision for manufacturing, one that he labels *dynamic evolution*. It requires greater worker autonomy and a commitment to building the problemsolving skills of the workforce. There is a suggestion that corporate culture—or at least shared goals—is necessary to obtain a high level of motivation and a congruence in goals between workforce and management. Quality circles are one vehicle for focusing, developing, and applying worker problemsolving skills.

Wheelwright recognizes that it is middle management that is most likely to feel threatened by such developments. Certain union structures are also threatened. One should not underestimate either the importance of these changes or the difficulty in getting them accomplished. Once accomplished, however, they can be the source of a sustained competitive advantage.

The progressive view sees manufacturing as a team sport. It involves a group of peers who work as equals across functions throughout the duration of the product development/manufacturing start-up activities. It also indicates that superior performance with respect to manufacturing process innovation involves more than hardware; it comprises the systems, people, equipment, and all aspects of an organization's production capabilities. Supporting evidence of the efficacy of the progressive approach can be gleaned from economic history, as well as contemporary cases such as General Electric Dishwasher and Mitsubishi Automotive Australia.

Lawrence (Chapter 5)

This chapter is designed to assist policymakers by describing how competitive forces, over time, have made U.S. industries and companies economically successful. In examining what is labelled the *social psychology of competition*, Lawrence arrives at the stark conclusion that an industry needs to experience vigorous competition if it is to be economically strong. Either too much or too little competitive pressure can lead an industry to a predictably weak economic performance. There appears to be a role for government, according to Lawrence, in moving selected industries away from unhealthy extremes of weak and cutthroat competition.

The history of the steel industry is used to exemplify the problems of insufficient competition. For seventy-five years after the foundation of U.S. Steel in 1901, the industry was characterized by relatively weak competition. It took the emergence of strong foreign competition in the 1970s to point up the weaknesses of the U.S. industry. The story of the auto industry from the emergence of the big three in the mid-1920s to the early 1970s follows similar lines.

The effects of too much competition have been similarly far reaching and destructive, though the effects are not widely recognized. Although the effects on each firm are different, the impact on the wider economy is similar. U.S. farming during the 1920s is cited as the classic example. The signs of excessive competition included a high rate of bankruptcies, wildly fluctuating pricing, limited gains in productivity, overcapacity, and a low rate of investment in readily available productivity-enhancing equipment. All of this occurred while the general economy was booming. A renewal was triggered in the 1930s, partly as a result of government price stabilization.

An important reason that excessive competition creates inefficiencies relates to the adjustment process. The neoclassical economic paradigm holds that the least efficient capacity will exist first when an industry is under competitive pressure. But when multiplant firms exist, Lawrence argues, management will shift production to more efficient plants if they have excess capacity. Single-plant firms that were more efficient than the ideal capacity might stay open even if they were less efficient. The point is (and there is supporting evidence available from Baden Fuller's work in Europe) that the least efficient capacity is not necessarily the first to exit. Further, the social psychology of severe scarcity can lead to many false economies such as the neglect of routine preventive maintenance—witness several U.S. airlines' maintenance violations during tough times in the mid-1980s. This leads Lawrence, based on his understanding of the social psychology of organizations, to appeal to economists to reactivate the concept of cutthroat competition that has somehow disappeared from the literature.

Innovation can similarly be impaired by change in the external environment that is too rapid as it leads to information overload. Uncertainty can also be disfunctional because it may obscure technological choices and render new investments risky.

Although government is only one factor that can influence the degree of competitive pressure in a given industry, it is often the most critical. Instruments at the government's disposal include antitrust, tax policy, trade policy, price supports, and price regulations. These instruments are already extensively utilized, but not in a coordinated fashion and not with superior industry performance as the goal. This leads Lawrence to propose a new government agency to monitor competitive pressure. It would intervene to deregulate or break up collusion when inefficient competition was observed; when excessive competition was observed, the government would intervene with trade protection, faster depreciation allowances, loan guarantees, and the like. Although the hazards of government intervention are recognized, Lawrence believes that government application of the competitive principle would boost business performance. With consistent and understandable rules, business could plan confidently for the long run and thereby attain superior performance.

Pfeffer (Chapter 6)

In this chapter, Pfeffer critiques methodological individualism—an approach to analyzing action that is central to economics but is also too often found in many other fields, including strategic management. With respect to the latter, both corporate strategy (which focuses on what markets or businesses the firm should be in) and business-level strategy (which focuses on competition within particular product/market segments) are characterized by a fundamentally internal focus with the single organization as the unit of analysis. Although it is true that both research

and practice in strategic management consider the nature and characteristics of the environment facing the firm, such as the number of competitors, market shares, and so forth, it is also true that in most research and practice these external constraints are taken as given, as characteristics of the environment to which firms must adapt in order to be successful. Organizations tend to be viewed as solitary units confronted by faceless environments. Strategic action thus becomes concerned with matching organizational capacities to environmental characteristics. Even when the focus is on an entire population (as with the population ecology approach) or an industry (as with industry analysis as taught at the Harvard Business School and elsewhere), the embedded, situational character of relationships among units is neglected. Environments tend to be characterized in terms of niches, resource pools, and the like. Competitors are faceless, and their actions are often not tied to the actions of the focal organization.[5] Pfeffer notes that systems of organizations are structured, as are organizations themselves, and there are institutional elements of these structures that ought to be recognized in research and practice. For instance, in the Freeman and Hanan study of restaurants, a given restaurant's likelihood of survival was predicted by its form (specialism and generalism) interacting with the conditions of the environment. No consideration is built into the analysis of the restaurant's patterns of relations with suppliers, customers, or competitors in the area. If profits, and possibly survival and stability, are related to interorganizational power, then one important objective of strategic management should be to enhance the firm's interorganizational power. Accepting this premise leads one further into considering the relational aspects of organizational life. Hence Pfeffer's insistence that interorganizational power[6] needs to be a key part of strategic management.

Pfeffer outlines several paradigms within which power has been considered. The structural autonomy approach sees power occurring to those actors who occupy positions with other actors that are centralized or coordinated, and deal with a diversity of other sectors that are themselves uncentralized and unable to engage in coordinated action. Thus, diversification is significant not because of possible economies or risk reduction attributes but because it reduces dependence on less concentrated and more dispersed customers. Diversification that causes the firm to depend on less concentrated and more dispersed sources of supply can be similarly understood. Likewise, coordination such as through trade associations, interlocking directorates, and joint ventures can be seen as an effort to enhance power.

The network analysis approach to power predicts that actors that are more centrally located and more interconnected should have more power. According to this approach, which in some ways is diametrically opposite to the structural autonomy approach, firms or sectors that are more central in exchange relations should have more power.

A third approach is resource dependence, which specifies conditions that would affect whether or not a given organization would comply with external demands and, by extension, its degree of autonomy and power. These include the need for the resource alternatives available and the focal organization's desire to survive.

Important implications for strategic management follow. Deeper appreciation than is currently evidenced is needed for the concept of collective strategy—that is, joint action by organizations on matters of strategic importance. Individual strategies can be overwhelmed by cooperative agreements such as joint ventures, strategic alliances, cooperative R&D, and the like. Put differently, collaborative arrangements can constitute what Richard Rumelt has called isolating mechanisms—factors that protect an individual firm's profit position and thereby help explain intraindustry differences in performance. There is no doubt that the profitability of the U.S. auto industry from the early 1980s on was due largely to intergovernmental agreements that limited Japanese imports. Such restrictions were achieved by a political coalition (collective action) involving both the companies and the unions.

The general thrust of the chapter is that the strategy literature needs to attend to a wider set of strategic actions and responses than it has in the past. Such activities include not only vertical and horizontal mergers but also joint ventures, board-of-director interlocks, and political activity of all shapes and varieties. This suggests the importance of achieving a merger between the subject-matter areas of strategy and business and public policy (or business and society) as they commonly exist in U.S. business schools. If power can be channeled by interfirm organization, greater attention needs to be given to interfirm coordination and organization. This ought to lead to a reconsideration of at least a refinement of the entry-barrier concept. Entry conditions are obviously conditioned by preexisting interfirm relations, as Robert Eccles's study of the U.S. construction industry and Michael Gerlach's study of Japanese Kereitsu (industrial groups) make starkly apparent.

A deeper understanding is also needed of how interfirm and industry level organizations develop and evolve. Studies of regulatory agencies such as the FCC, ICC, and CAB are instructive in this regard. But our understanding of how interfirm organizations affect coordination and outcomes is still quite rudimentary. Pfeffer forcefully and convincingly argues that the analysis of strategy and organization must be inextricably linked to the study of public policy. Put differently, the relationships of economic organizations to the state and a better understanding of the institutional factors of political life are necessary for the further development of the theory and practice of strategic management.

Important implications follow for the kinds of skills that ought to be possessed by those responsible for the foundation and implementation of strategy. If interorganizational linkages are important, then so are the political and organizational skills needed to develop them. It becomes necessary to identify and build coalitions of support, to organize and mobilize sometimes diverse interests, and to be sensitive to external, collective factors that affect the organization's well-being. Likewise, the success of top management ought to be seen as at least partly the consequence of the development and exercise of interorganizational skills. Implementation skills also must extend across organizations, as well as within the focal organization.

The conceptualization of business strategy advanced in this chapter clearly broadens the mainstream view of strategic management. The focus is outside the organization—on the structural and relational elements of societies and economies and on the fundamentally political nature of organizational strategy.

Rumelt (Chapter 7)

A basic problem with much of strategic management research is that it leaves out the entrepreneurial creation of new technical and commercial arrangements. This is particularly true with respect to "fit" theories, which involve matching the firm to its environment. Strategy theorists rely on fit theories in recommending adaptive moves; in certain ecological approaches, managers may be seen as simply the custodians of structures inherited from the past. Rumelt indicates that strategy researchers need to create, develop, and test a theory of entrepreneurship in order to remedy this situation.

A good working theory of entrepreneurship would not be tantamount to laying out an algorithm to get rich; rather, it would explain the conditions under which entrepreneurial talents would (and should) be employed. It would deal with the factors governing the supply and demand of entrepreneurial activity, with what resources need be associated with it and with the types of structural and contractual arrangements that are appropriate for entrepreneurial endeavor.

The objective of the chapter is to lay the groundwork for such a theory. The author's starting point is the notion that entrepreneurship involves the creation of new production functions; therefore, the incentives to innovate depend on the uncertainty of discovering a valuable asset or idea and the degree to which the social value it creates can be captured as private gain.

In order to build a theory of entrepreneurship, Rumelt begins by review-ing the relationship between economic rents and business strategy. In mainstream (textbook) microeconomic theory, profit-maximizing firms operating in competitive environments earn zero economic profits—that is, they are just able to cover their full costs. This has led many strategy researchers to turn to monopoly power arguments to explain persistent superior returns. Hence, firms earn surplus profits by colluding behind strategically erected entry barriers. Entry barriers are themselves in-sufficient because without collusion competition among firms behind the barriers would erode profits. A major empirical problem with this theory is that it cannot explain interfirm and intraindustry differences in pro-fitability, which empirically are much larger than interindustry differ-ences. The evidence strongly suggests that important sources of excess profitability are firm specific rather than the result of industry membership.

One response to this difficulty has been to redefine the unit of analysis to be strategic groups. An alternative approach, which Rumelt has pioneered, is to assert the presence of resource heterogeneity at the level of the firm. As a result, concepts of market power, monopoly, and oligopoly are replaced by a focus on unique firm-specific resources, limits on imitability, and on patterns of long-term contracting. Once the source of high profits is located in the firm's resource bundle rather than in its membership in a collective, the appropriate profit concept is that of rent. Rents, unlike profits, persist in equilibrium. The neoclassical dodge that preserves the zero-profit assumption is to ascribe the rents fully to the scarce factor and then to treat that factor as separately owned, so that the firm's costs include the rent on the factor. If the scarce factor is then traded, the rents are capitalized and no one (except some original owner) shows any profit.

The traditional foundation, Rumelt persuasively argues, is inadequate in the face of new insights. In particular, recent research on transactions cost economics highlights the fact that resources that can just as well be rented as owned are not all that common; they need to be generic assets for which the switching from one application to another does not involve a significant loss in value for the asset in question. If, however, as is often the case, the fixed rent-yielding factor is specialized to the needs of the firm, or if its use otherwise involves significant transactions costs, the rent on that factor is not logically or operationally separable from the profits of the firm.

Hence, Rumelt is led to make a distinction between Ricardian, Pare-tian, and entrepreneurial rents. The key to the existence of Ricardian rents is the presence of a fixed scarce factor. A standard way of presenting this notion is the increasing-cost industry. In this type of industry, it is possible to rank producers from least to highest cost, with the marginal cost of the least efficient producer equal to the market price. The marginal firm earns zero profit while the more efficient earns rents. The rent concept due to Pareto (and Marshall) is the difference between a resource's payment in its best use and the payment it would receive on its next best use. Thus, the Pareto rent is the payment received above and beyond that amount required to call it into use. When resources in use all have the same value in their best alternative use, the Ricardian and Pareto concepts correspond.

Whereas the classical concepts of rent apply in a static world and focus on the productivity of different resources or of resources in different uses, entrepreneurial rent stems from uncertainty and the discovery of new products and ways of doing business. Rumelt defines *entrepreneurial rent* as the difference between a venture's *ex post* value and the *ex ante* cost of the resources combined to form the venture. If one posits expectational equilibrium (*ex ante* cost equals expected *ex post value*), then expected entrepreneurial rents are zero. Hence, entrepreneurial rent springs from ex ante uncertainty. Ex post entrepreneurial rents are appropriable not by competitors but by the factors of production—such as the chef at a successful restaurant can attempt to recontract if the restaurant turns out to be a bonanza.

Given expectational equilibrium, it is uncertainty that produces the possibility of entrepreneurial rents. Absent uncertainty, one would expect the inputs used in an entrepreneurial venture to reflect their value in use. Otherwise, ex ante crowding or rapid imitation would reduce profits to normal levels. The resolution of this uncertainty is normally viewed as discovery or invention.

Entrepreneurial discovery includes technological innovation, real estate development, and mineral exploration. It might also include the discovery of new demand patterns and consumer needs. Where entrepreneurial activities completely resolve the original uncertainty, the results achieved (absent secrecy) could be perfectly imitated. In this case, entrepreneurial rents can flow from the sale of property rights, such as patents and trade secrets. If, however, the venture leaves considerable residual uncertainty as to the reasons for the success, entrepreneurial return cannot be obtained from the sale of property rights that explain "how to do it" or simply give permission to use protected formulas and techniques.

A risky entrepreneurial venture will not necessarily yield entrepreneurial rents, even if it succeeds. The product must not only provide social value (such as be an efficient replacement for a substitute or create an entirely new market possibility) but also be immune to efforts by employees, suppliers, manufacturers, distributors, and so forth to siphon off profits (see Chapter 9) and be resistant to imitation. The last condition requires the existence of *isolating mechanisms*—a term that Rumelt has created to describe impediments to the ex post imitative dissipation of entrepreneurial rents. Property rights are one form of isolating mechanisms; but although the legal system provides the entrepreneur with property rights over discoveries of minerals, certain inventions, written material, and trademarks, no such protection exists for the vast bulk of business innovation. Fortunately, there are often first-mover advantages—such as information asymmetries, learning and scale economies, buyer switching costs, reputation, and channel access—that create enough of an imitation lag that entrepreneurial rents exist even when property rights are unavailable.

Having developed a theory of entrepreneurial rent, Rumelt explores the organizational context of entrepreneurial activity. He asks two fundamental questions: Which organizations will innovate, and when will innovation be carried out within new ventures rather than in existing firms? Addressing the first question requires consideration of what economists call cannibalism—the phenomenon by which an innovation by an established firm may curtail sales from existing products. Cannibalism is a well-recognized phenomenon in the literature. Answering the second requires exploration of certain comparative properties of large and small firms with respect to the sponsorship of risky projects. Rumelt explores a form of incentive failure that leads to corporate myopia when management is mobile.

In large organizations, investment authorization is often made by managers laboring under considerable ignorance as to the true merits of projects. Top managers, lacking direct knowledge themselves, must rely on their subordinates' recommendations. Although this can be and is routinely probed, acceptance or otherwise depends on the proposing manager's track record. If such managers have considerable mobility, however, they are likely to temper their recommendations with the possibility that they will have left the organization by the time performance results are available. The net effect can be that mobile managers will discount future cash flows more heavily than would be indicated by their personal discount rates or their employers' cost of capital. The corporation as a whole will thus appear more myopic than are its members. (Capital market takeover cannot solve the problem because it is not top management that is at issue.) Top management will not be ignorant of this process, but it cannot identify which projects are opportunistically presented. It is forced to discount all claims about future profits even more sharply, thereby deepening institutional myopia.

This process creates obvious problems for the entrepreneur with the truly superior project. Top management will rationally discount the longer-time horizon proposals; seeing that the project will be rejected, the entrepreneurial manager has the incentive to leave the firm and pursue the project independently. This deepens myopia in the large enterprise because top management asks the nagging question of why someone has not already walked with the project if it is any good! This theory explains institutional myopia when all actors are rational and also explains employee exits and spinoffs in terms of incentive failure rather than intellectual theft.

Winter (Chapter 8)

Profound ambiguities often exist in both principle and practice regarding the scope and locus of the rights associated with the possession of knowledge assets. Sometimes it is well-defined intellectual property, like a patent; often it is skills, knowhow, competence, capability, or similar amorphous assets. Although the economics, finance, and accounting disciplines treat the asset as if it were well defined, it is commonly not, so that the tools of the basic disciplines contain little that afford a useful analytical grasp on strategy issues. Chapter 8 attempts to bridge the gap between the two meanings of *asset* as they relate to the knowledge and competence of the business firm.

Winter proposes viewing organizational strategy as a summary account of the principal characteristics and relationships of the organization and its environment, an account developed for the purpose of informing decisions affecting the organization's success and survival. This formulation emphasizes the normative content of strategic analysis and rejects the notion that there are strategies that have evolved implicitly or that strategy is a nonrational concept. Rather, Winter prefers to view strategy analysis as intensely rational behavior directed toward pragmatically useful understanding of the situation of the organization as a whole. Mere habits of thought or action, managerial or otherwise, are not strategies. Winter recognizes, however, that organizational repertories may be deeply ingrained so that the habits and impulse of an organization may be strong enough to severely delimit if not eliminate strategic choices.

A key step in strategic thinking must be the identification of the attributes of the organization that are considered subject to directed change. To identify these conceptually, Winter suggests utilizing the internal logic of control theory, where *state variables* are differentiated from *control* variables. The former are not subject to choice in the short run;[7] the latter are. The values chosen for control variables do, however, affect the evolution of the state variables over longer time spans. Winter advocates the development of a strategically useful characterization of those features of the organization that are not subject to choice in the short run but are influential in the long run and are considered key to its development and success.

If one then accepts present value as a goal, the logic of control theory can also provide guidance to the valuation of strategic assets. In particular it offers the full imputation principle, which states that a proper economic valuation of a collection of resources is one that precisely accounts for the returns the resources make possible. Owners of an asset should value it at the present value of the net future returns it generates under present ownership where the discount rate is set at an interest rate equal to the lending opportunities open to the owners. If more than this is offered, the owner of the asset should sell it. The adoption of this valuation principle implies that the notion that an excess return or (economic) profit can be earned by holding an asset is illusory. If there is gain or loss, the full imputation principle declares it to be capital gain or loss associated with having acquired the asset at a price below or above its true value. Such an outcome can be associated with good or bad luck and superior or inferior knowledge, competence, insight, skill, or information. The full imputation principle results in a conception of assets which is remote from financial accounting conventions. But Winter argues that the subtleties of the imputation dialectic—full imputation for one process discloses unaccounted returns in a casually antecedent process, which then calls for a fuller imputation—must be confronted if strategic analysis is to have solid foundations in economic reasoning. Whether confronted or not, they are the key issues when the strategic options include acceptance or rejection of a bid to purchase the company or part of it. Rational action in such situations demands that attention be paid to the question of the future earning power of the entity whose ownership is transferred. That question cannot be answered without inquiring deeply into the sources of earning power—that is, without confronting the imputation problem. The subtleties are particularly fundamental to understanding the strategic role of knowledge and competence.

In order to characterize approaches to a strategic problem where a conception of the decision variables is shared, Winter introduces the concept of a *heuristic frame*. Within a heuristic frame, there is room for a wide range of more specific formulations of a problem, but there is also enough structure provided by the frame itself to guide and focus discussions. A rich variety of different heuristic frames may nevertheless represent plausible approaches to a given problem. Commitment to a particular frame is thus a highly consequential step in strategic analysis.

Winter points out that many strategic perspectives can be recast into the language of state description and heuristic frames, often with the benefit of revealing gaps, limitations, or vagueness in the particular perspective. The danger of neglecting alternative heuristic frames can also be highlighted. To prove the point, Winter then spells out the heuristic frame of Boston Consulting Group's product portfolio approach and immediately identifies important gaps.

A description of knowledge states follows that is suggestive of heuristic frames appropriate to the development and protection of knowledge. Distinctions with respect to the protectability and transferability of codified and tacit knowledge are at the core. Winter suggests that such heuristic frames need to be applied to the learning curve concept. Mechanisms for skill accumulation and possession need to be understood—and they may differ across industries, technologies, and products—before the tool can be useful in strategic management. Evidence derived from the Yale survey of R&D executives is then presented. It establishes the great diversity of knowledge environments that exist in U.S. manufacturing—suggesting that heuristics derived from one industry may not be applicable in another.

Teece (Chapter 9)

This chapter attempts to explain why innovating firms often fail to obtain significant economic returns from an innovation, while customers, imitators, and other industry participants benefit. The phenomenon is well known to the British, whose pioneering innovations in aircraft and medicine have been copied and improved for decades. But it is also a phenomenon that innovative U.S. firms are confronting as fundamental developments and designs in industries such as microelectronics and aerospace are imitated and improved by both domestic and foreign competitors. A conceptual framework is developed that helps innovating firms derive market-entry strategies most likely to win. It indicates when an innovator can win big and explains when and why innovators will lose and therefore should not enter markets even though the innovation has real commercial value.

The most fundamental reason why innovators with good marketable ideas fail to enter or open up markets successfully is that they are operating in an environment where intellectual property is difficult to protect. The two most important environmental factors conditioning this are the efficacy of legal-protection mechanisms (such as patentability) and the nature of the technology (is it hard to copy, legal constraints aside?).

The design cycle—and the role that the innovator has in determining what will eventually emerge as the dominant design—also has critical impact on outcomes. With new product concepts, a great variety of designs may exist in the early stages. After considerable trial and error in the marketplace, one design or a narrow class of designs will emerge as the standard—such as the IBM PC in sixteen-bit personal computers, the Model T Ford in open body automobiles, the Douglas DC-3 in early passenger/cargo aircraft. The existence of a dominant design watershed is of great significance to the strategic positions of the innovator and its competitors as competition migrates from features toward price once the dominant design has emerged. Scale and learning become much more important. This typically requires industry participants to deploy specialized capital in order to obtain a cost advantage.

In most industries intellectual property protection is weak. Accordingly, it is important that the innovator settle on or establish the particular design that eventually will dominate. This is because there are substantial first-mover advantages and irreversibilities associated with tooling up for mass production. If follower/imitators, rather than the innovator, end up settling on what turns out to be the dominant design, then the innovator will be competitively disadvantaged, despite the fact that it pioneered. In these situations, the innovator ought to pursue multiple, parallel prototyping, if the cost penalties are not excessive. This will increase the chances of converting a lead in early commercialization into subsequent market-share primacy.

If an innovator is still in the game when the dominant design has emerged, then it must face a new problem—that of securing access to complementary assets. In almost all cases, the successful commercialization of an innovation requires that the know-how in question be utilized in conjunction with the services of other assets such as manufacturing, marketing, after sales services, and possibly other technologies. Often these services must be obtained from complementary assets that are specialized to the innovation—for example, VCRs cannot be made with all the same equipment used to make televisions. When this happens, it is difficult for the innovator to contract with others to provide these services because few firms will be willing to make irreversible investments in dedicated facilities to meet the needs of the innovator—without the innovator having a substantial share of the risk. It is widely recognized that prior ownership or subsequent integration into these complementary activities is needed if the required facilities are to be dedicated to the particular requirements of the purchaser. Hence, an innovator needs to have competitive capacities with respect to these complementary activities if it is to succeed in capturing value from its innovation. If imitators are better positioned with respect to the innovation's complementarities, then there is a high probability that imitators, or for that matter owners of cospecialized assets who joint-venture with imitators, will outcompete the innovator.

The framework suggests the importance of complementarities. That is why manufacturing matters. When intellectual property is difficult to protect with legal apparatus and when the innovation requires specialized skills and/or equipment in its manufacture, then the innovator probably will be unable to subcontract manufacturing and expect to win. Subcontracting would expose the innovator to possible recontracting hazards; in addition, manufacturing skills would tend to accumulate in the hands of the subcontractor not the innovator. Both features of the relationship expose the innovator to competitive threats by imitators and/or subcontractors with better access to competitive manufacturing.

These lessons are particularly pertinent to small, science-based companies. Founders with superb technical backgrounds are often slow to learn that generating advanced products and processes that meet a clear market need is not enough to guarantee success. Even with a terrific product, the innovator is likely to lose unless his intellectual property is extremely well protected or he is strategically well positioned with respect to key complementary assets. Needless to say, many science-based firms, after a shaky start, soon recognize these principles and adjust their strategies accordingly. Cetus Corporation in biotechnology is clearly playing by these rules, integrating into practically all of the specialized aspects of its core business, which it considers to be therapeutic cancer drugs.

There are powerful implications for nations that perceive their comparative advantage to be innovation. Unless one's trading partners are backward, it will never be enough to have the best science and engineering establishment and the most creative and innovative engineers and designers. Since the fruits of scientific effort are increasingly open to imitation because of weak intellectual property protection and low-cost technology transfer, extracting value from a nation's science and engineering prowess will require its firms to have competitive capacities in certain of the key complementary assets, such as manufacturing. In many cases this will need to be onshore in order to fashion defensible competitive strategies. Public policies that do not recognize that translating scientific and technological leadership into commercial leadership in most cases requires parallel excellence in capacities complementary to the innovation process will doom a nation to economic decline—possibly tempered only by a bounty of Nobel prizes. What *Business Week* calls hollow corporations will in general bring hollow success if the intended focus of their activities is innovation. Turning on both the private and public R&D spigots will not be enough for pioneering nations to regain their international competitiveness. It is the complementary capacities that must be built if a nation with the technological lead is also to lay claim to the commercial lead.

Weick (Chapter 10)

Weick argues that execution is analysis and that implementation is formulation. Strategy is thus an *ex post* construct; too much *ex ante* strategy can paralyze or splinter an organization. Strategy has its sources elsewhere—

in what Weick calls *strategy substitutes*. Strategy substitutes are conditions that neutralize what leaders do or that perform many leadership functions. They include characteristics of subordinates and characteristics of the organization—"state" variables in Winter's treatment in Chapter 8.

Some kind of plan or map may nevertheless be useful. Any old plan is often sufficient to get an organization moving; it is such movement that makes it possible to learn what is going on and what needs to be done next. A plan, even if vague, provides an excuse for people to act, learn, and create meaning. Action clarifies meaning. The pretext for action is less important than action itself. Strategic planning is but one of many pretexts for meaning-generation in organizations.

Action can thus substitute for strategy; strong beliefs that single out and intensify consistent action can bring events into existence. Whether people are fanatics, true believers, or idea champions, they all embody what looks like strategy in their persistent behavior. True believers impose their views on the world and fulfill their own prophecies. This makes strategy more of a motivational problem than a cognitive forecasting problem.

Presumptions can substitute for strategy. We often assume that people agree with us without ever testing that assumption. Vague strategic plans thus help because they allow the reality of disagreement to be skirted. The fact that disagreements persist undetected is not necessarily a problem because these differences provide a repertoire of beliefs and skills that allow an organization to cope with changing environments. Diverse skills and beliefs are an asset when environments are exposed to change.

Improvisation can also substitute for strategy. Strategic plans can become threats by restricting experimentation and the chance to learn that old assumptions no longer work. Elements of corporate culture such as logos and slogans can suffice to give general direction and shape improvisation in the hands of bright, ambitious, and confident people. Improvisation requires investment in general capabilities and a wide skill repertoire but compromises on bounded rationality by substituting adaptive capacities for a necessarily incomplete but burdensome plan. It is the essence of what Weick calls *just in time strategy*. Strategies are less accurately portrayed as episodes where people convene at one time to make a decision. They are more accurately portrayed as small steps (such as writing a memo or answering a query) that gradually foreclose alternative courses of action and limit what is possible. The crucial activities for strategy making are actions, the controlled execution of which consolidates fragments of policy that are lying around, gives them direction, and closes off other possible arrangements. Strategic planning is the pretext under which people act.

NOTES

1. The Bessemer process marked a radical departure from the small-scale batch process of the day and by breaking the constraints on speed and volume, laid the foundation for new markets and applications. New capabilities were essential to the success of the Bessemer process.
2. Continuous heat treating uses refinements and extensions of existing technology to create new market opportunities. Though it requires new equipment and a new approach to computer process control, it builds on established skills, knowledge, and experience found in a typical integrated steel mill but can create new customers through enhancing specialty offerings.
3. Continuous casting transforms liquid steel in semifinished slabs, billets, or blooms and eliminates the need to make ingots and then process them through a slabbing mill. Continuous casting builds on existing engineering skills, and higher quality and efficiency strengthens relationships with existing customers.
4. This technology would require the development of new knowledge, new equipment, new techniques, and new methods of process control and management. Because of the degree of automation, the workforce is likely to be largely made up of engineers and skilled technicians. The technology will be applied to existing markets.
5. This characterization is especially apt with respect to the population ecology non-approach to strategy; the industrial organization paradigm, while often viewing competitors as faceless, considers market responses through the concept of the *reaction function*. Furthermore, the best work explicitly takes into account how environments matter. Chapter 2, for instance, is centrally concerned with assessing how the international environment differs from the domestic and the implications for corporate strategy.
6. Those uncomfortable with the concept of power might choose to view power as the ability to appropriate what economists call *rents*.
7. Note that if evolutionary economics is used as the framework for strategy analysis, much of what management scientists consider as decision variables is considered to be given—governed by routines and not open to deliberate choice.

REFERENCES

President's Commission on Industrial Competitiveness. 1985. *Global Competition: The New Reality*, Vol. 1. Washington, D.C.: Superintendent of Documents.

U.S. Bureau of Labor Statistics. 1985. *Handbook of Labor Statistics*. Washington, D.C.: Superintendent of Documents.

Part I

American Industry in International Competition

2 CHANGING PATTERNS OF INTERNATIONAL COMPETITION

Michael E. Porter

Most observers conclude that international competition is high on the list of environmental changes facing firms today. The growing importance of international competition is well recognized both in the business and academic communities, for reasons that are fairly obvious when one looks at just about any data set that exists on international trade or investment. Figure 2–1, for example, compares world trade and world GNP. Something interesting started happening around the mid-1950s, when the growth in world trade began to significantly exceed the growth in world GNP. Foreign direct investment by firms in developing countries began to grow rapidly a few years later, about 1963 (United Nations Center on Transnational Corporations 1984). This period marked the beginning of a now widely recognized fundamental change in the international competitive environment. It is a trend that causes sleepless nights for many business managers.

A substantial literature on international competition exists because the subject is far from a new one. A large body of literature has investigated the many implications of the Heckscher-Ohlin model and other models of international trade, rooted in the principle of comparative advantage (see Caves and Jones 1985). The unit of analysis in this literature is the country. There is also considerable literature on the multinational firm, reflecting the growing importance of the multinational since the turn of the century. In examining the reasons for the multinational, much of this literature examines the multinational's ability to exploit intangible assets.[1] The work of Richard Caves and others has stressed the role of the multinational in transferring know-how and expertise gained in one country market to

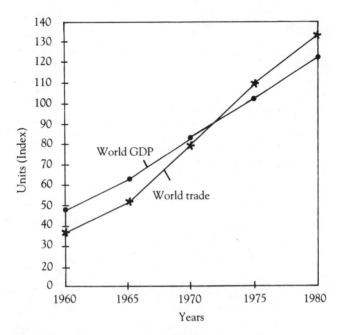

Source: United Nations, various years. *Statistical Yearbooks.*

Figure 2–1. Growth of World Trade.

others at low cost and thereby offsetting the unavoidable extra costs of doing business in a foreign country.

There is also a related literature on the problems of entry into foreign markets and the life cycle of how a firm competes abroad, beginning with export or licensing and ultimately moving to the establishment of foreign subsidiaries. Vernon's (1966) product cycle of international trade combines a view of how products mature with the evolution in a firm's international activities to predict the patterns of trade and investment in developed and developing countries.[2] Finally, many of the functional fields in business administration research have their branch of literature about international issues—such as international marketing and international finance. This literature concentrates, by and large, on the problems of doing business in a foreign country.

As rich as it is, however, the literature on international competition is limited when it comes to the choice of a firm's international strategy. Although the literature provides some guidance for considering incremental investment decisions to enter a new country, it provides at best a partial view of how to characterize a firm's overall international strategy and how such a strategy should be selected. The distinction between domestic firms and multinationals is seminal in a literature focused on the problems of doing business abroad, but the fact that a firm is multinational says little if anything about its international strategy except that it operates in several countries.

Broadly, my research has been seeking to answer the question, What does international competition mean for competitive strategy? In particular, what are the unique questions for competitive strategy that are raised by international as opposed to domestic competition? Many of the strategy issues for a company competing internationally are very much the same as for one competing domestically. A firm must still analyze its industry structure, understand its buyer and the sources of buyer value, diagnose its relative cost position, and seek to establish a sustainable competitive advantage within some competitive scope, whether it be across-the-board or in an industry segment. These are subjects that I have written about extensively (Porter 1980, 1985b). But there are some questions for strategy that are peculiar to international competition and that add to rather than replace those listed earlier. These questions all revolve, in one way or another, around how what a firm does in one country affects (or is what is affected by) what is going on in other countries—the connectedness among country competition. This connectedness is the focus of this chapter and of a broader stream of research recently conducted under the auspices of the Harvard Business School (see Porter, 1986).

Patterns of International Competition

The appropriate unit of analysis in setting international strategy is the industry because the industry is the arena in which competitive advantage is won or lost. The starting point for understanding international competition is the observation that its pattern differs markedly from industry to industry. At one end of the spectrum are industries that I call *multidomestic*, in which competition in each country (or small group of countries) is essentially independent of competition in other countries. A multidomestic industry is one that is present in many countries (for example, there is a consumer banking industry in Sri Lanka, one in France, and one in the United States) but in which competition occurs on a country-by-country basis. In a multidomestic industry, a multinational firm may enjoy a competitive advantage from the one-time transfer of know-how from its home base to foreign countries. However, the ongoing competitive outcome is determined by conditions in each country, and a multinational modifies and adapts its intangible assets to employ them in each country. The competitive advantages of the firm, then, are largely specific to each country. The international industry becomes a collection of essentially domestic industries—hence the term *multidomestic*. Industries where competition has traditionally exhibited this pattern include retailing, consumer packaged goods, distribution, insurance, consumer finance, and caustic chemicals.

At the other end of the spectrum are what I term *global* industries. The term *global*—like the word *strategy*—has become overused and perhaps underunderstood. The definition of a global industry employed here is an industry in which a firm's competitive position in one country is significantly impacted by its position in other countries.[3] Therefore, the international industry is not merely a collection of domestic industries but a series of linked domestic industries in which the rivals compete against each other on a truly worldwide basis. Industries exhibiting the global pattern today include commercial aircraft, television sets, semi-conductors, copiers, automobiles, and watches.

The implications for strategy of the distinction between multidomestic and global are profound. In a multidomestic industry, a firm can and should manage its international activities like a portfolio. Its subsidiaries or other operations around the world should each control all the important activities necessary to do business in the industry and should enjoy a high degree of autonomy. The firm's strategy in a country should be determined largely by the circumstances in that country; the firm's international strategy is then what I term a *country-centered strategy*.

In a multidomestic industry, competing internationally is discretionary. A firm can choose to remain domestic or can expand internationally if it has some advantage that allows it to overcome the extra costs of entering and competing in foreign markets. The important competitors in multidomestic industries will either be domestic companies or multinationals with stand-alone operations abroad—this is the situation in each of the multidomestic industries listed earlier. In a multidomestic industry, then, international strategy collapses to a series of domestic strategies. The issues that are uniquely international revolve around how to do business abroad, how to select good countries in which to compete (or assess country risk), and mechanisms to achieve the one-time transfer of know-how. These questions are relatively well developed in the literature.

In a global industry, however, managing international activities like a portfolio will undermine the possibility of achieving competitive advantage. In a global industry, a firm must in some way integrate its activities on a worldwide basis to capture the linkages among countries. This includes, but requires more than, transferring intangible assets among countries. A firm may choose to compete with a country-centered strategy, focusing on specific market segments or countries when it can carve out a niche by responding to whatever local country differences are present. However, it does so at some considerable risk from competitors with global strategies. All the important competitors in the global industries listed earlier compete worldwide with coordinated strategies.

In international competition a firm always has to perform some functions in each of the countries in which it competes. Even though a global competitor must view its international activities as an overall system, it must maintain some country perspective. The balancing of these two perspectives becomes one of the essential questions in global strategy.[4]

Causes of Globalization

If we accept the distinction between multidomestic and global industries as an important taxonomy of patterns of international competition, a number of crucial questions arise: When does an industry globalize? What exactly do we mean by a global strategy, and is there more than one kind? What determines the type of international strategy that should be selected in a particular industry?

An industry is global if there is some competitive advantage to integrating activities on a worldwide basis. To make this statement operational, however, we must be precise about what we mean by *activities* and *integrating*. To diagnose the sources of competitive advantage in any context, whether it be domestic or international, it is necessary to adopt a disaggregated view of the firm. In my newest book, *Competitive Advantage*, I have developed a framework for doing the so-called value chain, (Porter 1985b). Every firm is a collection of discrete activities performed to do business and occurring within the scope of the firm. The activities performed by a firm include such things as salespeople selling the product, service technicians performing repairs, scientists in the laboratory designing process techniques, and accountants keeping the books. Such activities are technologically and in most cases physically distinct. Only at the level of discrete activities, rather than the firm as a whole, can competitive advantage be understood.

A firm may possess two types of competitive advantage—low relative cost (its ability to perform the activities in its value chain at lower cost) or differentiation (performing in a unique way relative to its competitors). The ultimate value that a firm creates is what buyers are willing to pay for what the firm provides, which includes the physical product as well as any other services or benefits. Profit results if the value created through performing the required activities exceeds the collective cost of performing them. Competitive advantage is a function of either providing comparable buyer value to competitors but performing activities efficiently (low cost) or of performing activities at comparable cost but in unique ways that create greater buyer value than competitors and hence command a premium price (differentiation).

The value chain, shown in Figure 2–2, provides a systematic means of displaying and categorizing activities. The activities performed by a

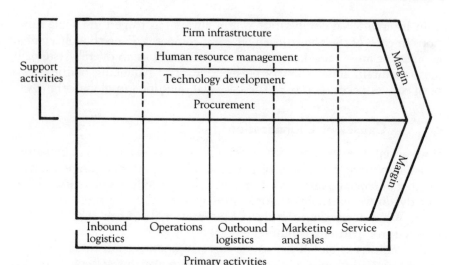

Figure 2–2.   The Value Chain.

firm in any industry can be grouped into the nine generic categories shown. The labels may differ based on industry convention, but every firm performs these basic categories of activities in some way or another. Within each category of activities, a firm typically performs a number of discrete activities that are particular to the industry and to the firm's strategy. In service, for example, firms typically perform such discrete activities as installation, repair, parts distribution, and upgrading.

The generic categories of activities can be grouped into two broad types: Along the bottom are what I call *primary* activities, which are those involved in the physical creation of the product or service, its delivery and marketing to the buyer, and its support after sale. Across the top are what I call *support* activities, which provide inputs or infrastructure that allow the primary activities to take place on an ongoing basis.

Procurement is the obtaining of purchased inputs, including raw materials, purchased services, and machinery. Procurement stretches across the entire value chain because it supports every activity—every activity uses purchased inputs of some kind. Typically many different discrete procurement activities occur within a firm, and often they are performed by different people. Technology development encompasses the activities involved in designing the product as well as in creating and improving the way the various activities in the value chain are performed. We tend to think of technology in terms of the product or manufacturing process. In fact, every activity a firm performs involves a technology or technologies, which may be mundane or sophisticated, and a firm has a stock of know-how about how to perform each activity. Technology development typically involves a variety of different discrete activities, some performed outside the R&D department.

Human resource management is the recruiting, training, and development of personnel. Every activity involves human resources, and human resource management activities cut across the entire chain. Finally, firm infrastructure includes activities such as general management, accounting, legal, finance, strategic planning, and all the other activities decoupled from specific primary or support activities but that are essential to enable the entire chain to operate.

Activities in a firm's value chain are not independent but are connected through what I call *linkages*. The way one activity is performed frequently affects the cost or effectiveness of other activities. If more is spent on the purchase of a raw material, for example, a firm may lower its cost of fabrication or assembly. There are many linkages that connect activities, not only within the firm but also with its suppliers, channels, and ultimately its buyers. The firm's value chain resides in a larger stream of activities that I term the *value system*. Suppliers have value chains that provide the purchased inputs to the firm's chain; channels have value chains through which the firm's product or service passes; buyers have value chains in which the firm's product or service is employed. The connections among activities in this vertical system also become essential to competitive advantage.

A final important building block in value chain theory, necessary for our purposes here, is the notion of *competitive scope*. Competitive scope is the breadth of activities that the firm employs when competing in an industry. There are four basic dimensions of competitive scope: (1) segment scope, or the range of segments that the firm serves (such as product varieties and customer types); (2) industry scope, or the range of industries that the firm competes in using a coordinated strategy; (3) vertical scope, or activities that are performed by the firm versus by suppliers and channels; and (4) geographic scope, or the geographic regions that the firm operates in using a coordinated strategy. Competitive scope is vital to competitive advantage because it shapes the configuration of the value chain, how activities are performed, and whether activities are shared among units. International strategy is an issue of geographic scope and can be analyzed quite similarly to the question of whether and how a firm should compete locally, regionally, or nationally within a country. In the international context, government tends to have a greater involvement in competition and there are more significant variations among geographic in buyer needs, although these differences are matters of degree.

International Configuration and Coordination of Activities

A firm that competes internationally must decide how to spread the activities in the value chain among countries. A distinction immediately arises between the activities labeled downstream on Figure 2–3 and those labeled upstream activities and support activities. The location of downstream activities, those more related to the buyer, is usually tied to where the buyer is located. If a firm is going to sell in Japan, for example, it usually must provide service in Japan and must have salespeople stationed in Japan. In a few industries it is possible to have a single salesforce that travels to the buyer's country and back again; some other specific downstream activities such as the production of advertising copy also sometimes can be done centrally. More typically, however, the firm must locate the capacity to perform downstream activities in each of the countries in which it operates. Conversely, upstream activities and support activities at least conceptually can be decoupled from where the buyer is located.

This distinction carries some interesting implications. The first is that downstream activities create competitive advantages that are largely country-specific: A firm's reputation, brand name, and service network in a country grow out of a firm's activities in that country and create entry/mobility barriers largely in that country alone. Competitive advantage in upstream and support activities, however, often grows more out of the entire system of countries in which a firm competes than from its position in any one country.

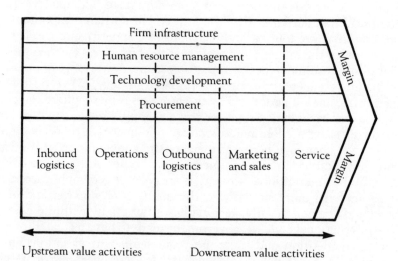

Figure 2–3.   Upstream and Downstream Activities.

A second implication is that in industries where downstream activities or buyer-tied activities are vital to competitive advantage, there tends to be a more multidomestic pattern of international competition. In industries where upstream and support activities such as technology development and operations are crucial to competitive advantage, global competition is more common. In global competition, the location and scale of these potentially footloose activities is optimized from a worldwide perspective.[5]

The distinctive issues in international, as contrasted to domestic, strategy can be summarized in two key dimensions of how a firm competes internationally. The first is what I term the *configuration* of a firm's activities worldwide—or where in the world each activity in the value chain is performed, including in how many places. The second dimension is what I term *coordination*—which refers to how like activities performed in different countries are coordinated with each other. If, for example, there are three plants—one in Germany, one in Japan, and one in the United States—how do the activities in those plants relate to each other?

A firm faces an array of options in both configuration and coordination for each activity. Configuration options range from concentrated (performing an activity in one location and serving the world from it, such as one R&D lab or one large plant) to dispersed (performing every activity in each country). In the latter case, each country would have a complete value chain. Coordination options range from none to very high. If a firm produces its product in three plants, for example, it could at one extreme allow each plant to operate with full autonomy, such as different product standards and features, different steps in the production process, different raw materials, different part numbers. At the other extreme, the plants could be tightly coordinated by employing the same information system, the same production process, the same parts, and so forth. Options for coordination in an activity are typically more numerous than the configuration options because there are many possible levels of coordination and many different facets of the way the activity is performed.

Table 2–1 lists some of the configuration issues and coordination issues for several important categories of value activities. In technology development, for example, the configuration issue is where R&D is performed: one location or two locations and in what countries? The coordination issues have to do with such things as the extent of interchange among R&D centers and the location and sequence of product introduction around the world. There are configuration issues and coordination issues for every activity.

Figure 2–4 is a way of summarizing these basic choices in international strategy on a single diagram, with coordination of activities on the vertical axis and configuration of activities on the horizontal axis. The firm has to

Table 2–1. Configuration and Coordination Issues by Category of Activity

Value Activity	Configuration Issues	Coordination Issues
Operations	Location of production facilities for components and end products	Networking of international plants Transferring process technology and production know-how among plants
Marketing and sales	Product line selection Country (market) selection	Commonality of brand name worldwide Coordination of sales to multinational accounts Similarity of channels and product positioning worldwide Coordination of pricing in different countries
Service	Location of service organization	Similarity of service standards and procedures worldwide
Technology development	Number and location of R&D centers	Interchange among dispersed R&D centers Developing products responsive to market needs in many countries Sequence of product introductions around the world
Procurement	Location of the purchasing function	Managing suppliers located in different countries Transferring market knowledge and Coordinating purchases of common items

make a set of choices for each activity. If a firm employs a very dispersed configuration, placing an entire value chain in every country (or small group of contiguous countries) in which it operates, coordinating little or not at all among them, then the firm is competing with a country-centered strategy. The domestic firm that operates in only one country is the extreme case of a firm with a country-centered strategy. As we move from the lower left-hand corner of the diagram up or to the right, we have strategies that are increasingly global.

Figure 2–5 illustrates some of the possible variations in international strategy. The purest global strategy is to concentrate as many activities as possible in one country, serve the world from this home base, and tightly coordinate those activities that must inherently be performed near the buyer. This is the pattern adopted in the 1960s and 1970s by many Japanese firms, such as Toyota. However, Figures 2–4 and 2–5 make it

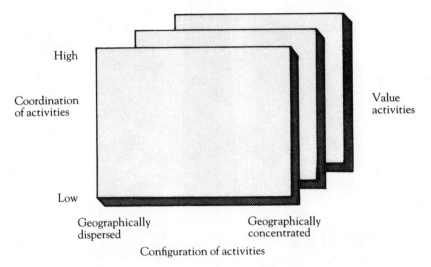

Figure 2–4. The Dimensions of International Strategy.

clear that there is no such thing as one global strategy. There are many different kinds of global strategies, depending on a firm's choices about configuration and coordination throughout the value chain. In copiers, for example, Xerox has until recently concentrated R&D in the United States but dispersed other activities, in some cases using joint-venture partners to perform them. On dispersed activities, however, coordination has been quite high. The Xerox brand, marketing approach, and servicing procedures have been standardized worldwide. Canon, on the other hand, has had a much more concentrated configuration of activities and somewhat less coordination of dispersed activities. The vast majority of support activities and manufacturing of copiers have been performed in Japan. Aside from using the Canon brand, however, local marketing subsidiaries have been given quite a bit of latitude in each region of the world.

A global strategy can now be defined more precisely as one in which a firm seeks to gain competitive advantage from its international presence through either configuration or coordination or both. Measuring the presence of a global industry empirically must reflect both dimensions and not just one. Market presence in many countries and some export and import of components and end products are characteristic of most global industries. High levels of foreign investment or the mere presence of multinational firms are not reliable measures, however, because firms may be managing foreign units like a portfolio.

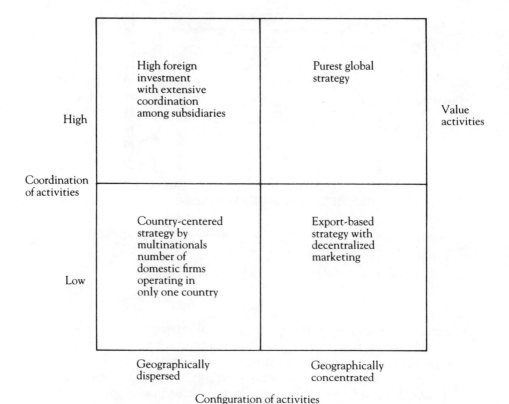

Figure 2–5. Types of International Strategy.

Configuration/Coordination and Competitive Advantage

Understanding the competitive advantages of a global strategy and, in turn, the causes of industry globalization requires specifying the conditions in which concentrating activities globally and coordinating dispersed activities leads to either cost advantage or differentiation. In each case, there are structural characteristics of an industry that work for and against globalization.

The factors that favor concentrating an activity in one or a few locations to serve the world are as follows: economies of scale in the activity; a proprietary learning curve in the activity; comparative advantage in where the activity is performed; and coordination advantages of co-locating linked activities such as R&D and production. The first two factors relate to *how many* sites an activity is performed at, while the last two relate to *where* these sites are. Comparative advantage can apply to any activity,

not just production. There may be some locations in the world that are
better places than others to do research on medical technology or to
perform software development, for example. Government can promote
the concentration of activities by providing subsidies or other incentives
to use a particular country as an export base—in effect altering comparative advantage—a role that many governments are playing today.

There are also structural characteristics that favor dispersion of an
activity to many countries, which represent concentration costs. Local
product needs may differ, nullifying the advantages of scale or learning
from one-site operation of an activity. Transport, communication, and
storage costs may make it inefficient to concentrate the activity in one
location. Government is also frequently a powerful force for dispersing
activities. Governments typically want firms to locate the entire value
chain in their country because this creates benefits and spillovers to the
country that often go beyond local content. Dispersion is also encouraged
by the risks of performing an activity at one place: exchange rate risks,
political risks, and so on. The balance between the advantages of concentrating and dispersing an activity normally differ for each activity (and
industry). The best configuration for R&D is different than that for
component fabrication, and this is different from that for assembly, installation, advertising, and procurement.[6]

The desirability of coordinating like activities that are dispersed involves a similar balance of structural factors. Coordination potentially
allows the sharing of know-how among dispersed activities. If a firm
learns how to operate the production process better in Germany, transferring that learning may make the process run better in the United
States and Japanese plants. Coordination potentially improves the ability
to reap economies of scale in activities if subtasks are allocated among
locations to allow some specialization—each R&D center has a different
area of focus. Although there is a fine line between such forms of coordination and what I have termed *configuration*, it illustrates how the way
that a network of foreign locations is managed can have a great influence
on the ability to reap the benefits of any given configuration of activities.
Viewed another way, close coordination is frequently a partial offset to
dispersing an activity.

Coordination may also allow a firm to respond to shifting comparative
advantage, where shifts in exchange rates and factor costs are hard to
forecast. Incrementally increasing the production volume at the location
currently enjoying favorable exchange rates, for example, can lower overall costs. Coordination can reinforce a firm's brand reputation with buyers,
and hence differentiation, through ensuring a consistent image and
approach to doing business on a worldwide basis. This is particularly likely
if buyers are mobile or if information about the industry flows freely

around the world. Coordination may also differentiate the firm with multinational buyers if it allows the firm to serve them anywhere and in a consistent way. Coordination (and a global approach to configuration) enhances leverage with local governments if the firm is able to grow or shrink activities in one country at the expense of others. Finally, coordination yields flexibility in responding to competitors, by allowing the firm to differentially respond across countries and to respond in one country to a challenge in another.

Coordination of dispersed activities usually involves costs that differ by form of coordination and industry. Local conditions may vary in ways that may make a common approach across countries suboptimal. If every plant in the world is required to use the same raw material, for example, the firm pays a penalty in countries where that raw material is expensive relative to satisfactory substitutes. Business practices, marketing systems, raw material sources, local infrastructures, and a variety of other factors may differ across countries as well, in ways that may mitigate the advantages of a common approach or of the sharing of learning. Governments may restrain the flow of information required for coordination or may impose other barriers to it. Transaction costs of coordination, which have recently received increased attention in domestic competition, are vitally important in international strategy (see Williamson 1975; Teece 1986; Casson 1982). International coordination involves long distances, language problems, and cultural barriers to communication. These may mean in some industries that coordination is not optimal. They also suggest that forms of coordination that involve relatively infrequent decisions will enjoy advantages over forms of coordination involving ongoing interchange.

Substantial organizational difficulties also are involved in achieving cooperation among subsidiaries because of difficulties in aligning subsidiary managers' interests with those of the firm as a whole. The Germans do not necessarily want to tell the Americans about their latest breakthroughs on the production line because it may make it harder for them to outdo the Americans in the annual comparison of operating efficiency among plants. These vexing organizational problems mean that country subsidiaries often view each other more as competitors than collaborators.[7] As with configuration, a firm must make an activity-by-activity choice about where there is net competitive advantage from coordinating in various ways.

Coordination in some activities may be necessary to reap the advantages of configuration in others. The use of common raw materials in each plant, for example, allows worldwide purchasing. Moreover, tailoring some activities to countries may allow concentration and standardization of others. For example, tailored marketing in each country may allow the same product to be positioned differently and hence sold successfully in many countries, unlocking possibilities for reaping economics or scale in production and R&D. Thus coordination and configuration interact.

When benefits of configuring and/or coordinating globally exceed the costs, an industry will globalize in a way that reflects the net benefits by value activity. The activities in which global competitors gain competitive advantage will differ correspondingly. An industry such as commercial aircraft represents an extreme case of a global industry, in the upper right-hand corner of Figure 2–5. There are three competitors, all with global strategies: Boeing, McDonnell Douglas, and Airbus. In activities important to cost and differentiation in the industry, there are compelling net advantages to concentrating most activities and coordinating the dispersed activities extensively.[8] In R&D there is a large fixed cost of developing an aircraft model ($1 billion or more) that requires worldwide sales to amortize. There are significant economies of scale in production, a steep learning curve in assembly (the learning curve was born out of research in this industry), and apparently significant advantages of locating R&D and production together. Sales of commercial aircraft are infrequent via a highly skilled salesforce, so that even the salesforce can be partially concentrated in the home country and travel to buyers.

The costs of a concentrated configuration are relatively low in commercial aircraft. Product needs are homogeneous, and there are the low transport costs of delivering the product to the buyer. Worldwide coordination of the one important dispersed activity, service, is very important—standardized parts and repair advice have to be available wherever the plane lands.

As in every industry, there are structural features that work against a global strategy in commercial aircraft. These are all related to government, a not atypical circumstance. Government has a particular interest in commercial aircraft because of its large trade potential, the technological sophistication of the industry, its spillover effects to other industries, and its implications for national defense. Government also has an unusual degree of leverage in the industry: In many instances it is the buyer. Many airlines are government owned, and a government official or appointee is head of the airline.

The competitive advantages of a global strategy is so great that all the successful aircraft producers have sought to achieve and preserve them. In addition, the power of government to intervene has been mitigated by the few viable worldwide competitors and the enormous barriers to entry created in part by the advantages of a global competitor. The result has been that firms have to assuage government through procurement. Boeing, for example, is very careful about where it buys components. In countries that are large potential customers, Boeing seeks to develop suppliers. This requires a great deal of extra effort by Boeing to transfer technology and to work with suppliers to assure that they meet its standards. Boeing realizes that this is preferable to compromising the competitive advantage of its strongly integrated worldwide strategy. It is willing to employ one value activity (procurement) where the advantages of concentration are modest to help preserve the benefits of concentration in other activities. Recently, commercial aircraft competitors have entered into joint ventures and other coalition arrangements with foreign suppliers to achieve the same effect, as well as to spread the risk of huge development costs.

The extent and location of advantages from a global strategy vary among industries. In some industries, the competitive advantage from a global strategy comes in technology development, although firms gain little advantage in the primary activities so that these are dispersed around the world to minimize concentration costs. In other industries such as cameras or videocassette recorders, a firm cannot succeed without concentrating production to achieve economies of scale but gives subsidiaries much local autonomy in sales and marketing. In some industries, finally, there is no net advantage to a global strategy and country-centered strategies dominate—the industry is multidomestic.

Segments of an industry frequently vary in their pattern of globalization. In aluminum, the upstream (alumina and ingot) stages of the industry are global businesses. The downstream stage, semifabrication, is a group of country-centered businesses because product needs vary by country, transport costs are high, and intensive local customer service is required. Scale economies in the value chain are modest. In motor oil, automotive oil tends to be a country-centered business, while marine motor oil is a global business. In automotive oil, countries have varying driving standards, weather conditions, and local laws. Production involves blending various kinds of crude oils and additives and is subject to few economies of scale but high shipping costs. Country-centered competitors, such as Castrol and Quaker State, are leaders in most countries. In the marine segment, conversely, ships move freely around the world and require the same oil everywhere. Successful competitors are global.

The ultimate leaders in global industries are often first movers—the first firms to perceive the possibilities for a global strategy. Boeing was the first global competitor in aircraft, for example, as was Honda in motorcycles, and Becton Dickinson in disposable syringes. First movers gain scale and learning advantages that are difficult to overcome. First-mover effects are particularly important in global industries because of the association between globalization and economies of scale and learning achieved through worldwide configuration/coordination. Global leadership shifts if industry structural change provides opportunities for leapfrogging to new products or new technologies that nullify past leaders' scale and learning—again, the first mover to the new generation/technology often wins.

Global leaders often begin with some advantage at home, whether it be low labor cost or a product or marketing advantage. They use this as a lever to enter foreign markets. Once there, however, the global competitor converts the initial home advantage into competitive advantages that grow out of its overall worldwide system, such as production scale or ability to amortize R&D costs. Although the initial advantage may have been hard to sustain, the global strategy creates new advantages that can be much more durable.

International strategy has often been characterized as a choice between worldwide standardization and local tailoring or as the tension between the economic imperative (large-scale efficient facilities) and the political imperative (local content, local production). It should be clear from the discussion so far that neither characterization captures the richness of a firm's international strategy choices. A firm's choice of international strategy involves a search for competitive advantage from configuration/ coordination throughout the value chain. A firm may standardize (concentrate) some activities and tailor (disperse) others. It may also be able to standardize and tailor at the same time through the coordination of dispersed activities or use local tailoring of some activities (such as different product positioning in each country) to allow standardization of others (such as production). Similarly, the economic imperative is not always for a global strategy—in some industries a country-centered strategy is the economic imperative. Conversely, the political imperative is to concentrate activities in some industries where governments provide strong export incentives and locational subsidies.

Global Strategy and Comparative Advantage

The traditional concept of comparative advantage is that factor-cost or factor-quality differences among countries lead to production in countries with advantages in a particular industry that export the product elsewhere in the world. Competitive advantage, in this view, grows out of *where* firms perform activities.

The location of activities is clearly one source of potential advantage in a global firm. However, the global competitor can locate activities *wherever* comparative advantage lies, decoupling comparative advantage from the firm's home base or country of ownership. Indeed, the framework presented here suggests that comparative advantage not only involves production activities (the usual focus of discussions) but also applies to other activities in the value chain, such as R&D, processing orders, or designing advertisements.

Comparative advantage is specific to the *activity* and not the location of the value chain as a whole. One of the potent advantages of the global firm is that it can spread activities to reflect different preferred locations, something a domestic or country-centered competitor does not do. Thus, components can be made in Taiwan, software written in India, and basic R&D performed in Silicon Valley, for example. This international specialization and arbitrage of activities within the firm is made possible by the growing ability to coordinate and configure globally.

While my framework suggests a more complex view of comparative advantage, it also suggests that many forms of competitive advantage for the global firm derive less from *where* it performs activities than from *how* it performs them on a worldwide basis. Economies of scale, proprietary learning, and differentiation with multinational buyers are not tied to countries but to the configuration and coordination of the firm's worldwide system. Traditional sources of comparative advantage can be very elusive sources of competitive advantage for an international competitor today. Comparative advantage frequently shifts. A country with the lowest labor cost, for example, is overtaken within a few years by some other country as has happened repeatedly in the shipbuilding industry (Japan has replaced Europe only to be replaced by Korea).

Yet a global firm's overall system advantages—economies of scale, accumulated know-how, a worldwide brand reputation—are frequently much more sustainable advantages. Not only this, but the global competitor can, through locating activities abroad and coordination, nullify and even exploit shifts in comparative advantage that affect its home country. Many Japanese firms, for example, moved some activities years ago to Taiwan, Korea, and Hong Kong and solidified rather than harmed their positions.

My research on a broad cross-section of industries suggests that the achievement of sustainable world leadership follows a more complex pattern than the exploitation of comparative advantage *per se*. A competitor may start with a comparative-advantage–related edge that provides the basis for penetrating foreign markets, but this edge is rapidly translated into a broader array of advantages that arise from the global approach to competing. Over time, these systemwide advantages are further reinforced with country-specific advantages in *foreign* countries such as brand identity and relationships with local distribution channels.

Many Japanese firms were fortunate enough to make their transitions from country-based comparative advantage to global competitive advantage in a buoyant world economy while nobody paid much attention to them. European and U.S. competitors were willing to cede market share in "less desirable" segments such as the low end of the product line, or so they thought. The Japanese translated these beachheads into world leadership by broadening their lines and reaping advantages in scale and proprietary learning.

Global Platforms

The interaction of the home country conditions and the need and ability to create competitive advantages from a global strategy that transcends the country suggest a view of country success in industries that is more complex than comparative advantage. To understand this more complex role of the country, I define the concept of a *global platform*. A country is a desirable global platform if it provides an environment yielding firms domiciled in that country an advantage in competing globally in a particular industry.[9] An essential element of this definition is that it requires success *outside* the country and not merely country conditions that allow firms to successfully master domestic competition. There are three determinants of a good global platform in an industry, which I have explored in more detail elsewhere (Porter 1985a). The first is comparative advantage, in an extended sense that includes such things as access to low-cost imported inputs as well as cultural factors. The second determinant is country characteristics that lead firms domiciled there to be first in perceiving and choosing the appropriate global strategy—I term these *incubating conditions for first movers*. Such conditions confer on firms from a country a lead in capturing the scale and learning advantages of a global strategy. The third determinant of good global platform is country attributes that create *disproportionate home demand* for appropriate product varieties. These three determinants reflect the interaction between conditions of local supply, economies of scale and learning, and the composition and timing of country demand in shaping international success.

A good example of the interplay among these factors is the television set industry. In the United States early demand was in large screen console sets because television sets were initially luxury items kept in the living room. As buyers began to purchase second and third sets, sets became smaller and more portable. They were used increasingly in the bedroom, the kitchen, the car, and elsewhere. As the television set industry matured, table model and portable sets became the universal product variety. Japanese firms, because of the small size of Japanese homes, cut their teeth on small sets. They dedicated most of their R&D to developing small picture tubes and to making sets more compact. In the process of naturally serving the needs of their home market, then, Japanese firms gained early experience and scale in segments of the industry that came to dominate world demand. U.S. firms, conversely, cut their teeth on large-screen console sets with fine furniture cabinets. As the industry matured, the experience base of U.S. firms was in a segment that was small and isolated to a few countries, notably the United States. Japanese firms were able to penetrate world markets in a segment that was both uninteresting to foreign firms and one in which they had initial scale, learning, and labor cost advantages. Ultimately the low-cost advantage disappeared as production was automated, but global scale and learning economies took over as the Japanese advanced product and process technology at a rapid pace.

Three broad determinants of a good global platform rest on the interaction between country characteristics and firms' strategies. The literature on comparative advantage, through focusing on the country and suppressing the individual firm, is most appropriate in industries where there are few economies of scale, little proprietary technology or technological change, or no possibilities for product differentiation. Although these industry characteristics are those of many traditionally traded goods, they describe few of today's important global industries. Where it does recognize scale economies, trade theory views them narrowly as arising from production in one country.

The Evolution of International Competition

Having established a framework for understanding the globalization of industries, we are now in a position to view the phenomenon in historical perspective. If one goes back far enough, relatively few industries were global. Around 1880 most industries were local or regional in scope.[10] The reasons are rather self-evident in the context of our framework. There were few economies of scale in production until fuel-powered machines and assemblyline techniques emerged. There were heterogeneous product needs among regions within countries, much less among countries. There were few if any national media—the *Saturday Evening Post* was the first important national magazine in the United States and developed in the teens and twenties. Communicating between regions was difficult before the telegraph and telephone, and transportation was slow until the railroad system became well developed.

These structural conditions created little impetus for the widespread globalization of industry. Those industries that were global reflected classic comparative advantage considerations—goods were simply unavailable in some countries that imported them from others, or differences in the availability of land, resources, or skilled labor made some countries desirable suppliers to others. Export of local production was the form of global strategy adapted. There was little role or need for widespread government barriers to international trade during this period, although for centuries there have been trade disputes in limited product areas that had high political visibility.

Around the 1880s, however, were the beginnings of what today has blossomed into the globalization of many industries. The first wave of modern global competitors grew up in the late 1800s and early 1900s. Many industries went from local (or regional) to national in scope, and some began globalizing. Firms such as Ford, Singer, Gillette, National Cash Register, Otis, and Western Electric had commanding world market shares by the teens and operated with integrated worldwide strategies. Early global competitors were principally U.S. and European companies.

Driving this first wave of modern globalization were rising production scale economies due to advancements in technology that outpaced the growth of the world economy. Product needs also became more homogenized in different countries as knowledge and industrialization diffused. Transport improved, first through the railroad and steamships and later in trucking. Communication became easier with the telegraph and then the telephone. At the same time, trade barriers were initially modest.

The burst of globalization soon slowed, however. Most of the few industries that were global moved increasingly towards a multidomestic pattern—multinationals remained but between the 1920s and 1950 they often evolved toward federations of autonomous subsidiaries. The principal reason was a strong wave of nationalism and resulting high tariff barriers, partly caused by the world economic crisis and world wars. Another barrier to global strategies, chronicled by Chandler (1985), was a growing web of cartels and other interfirm contractual agreements. These limited the geographic spread of firms.

The early global competitors began rapidly dispersing their value chains. The situation of Ford Motor Company is no exception. In 1925 Ford had almost no production outside the United States, but by World War II its overseas production had risen sharply. Firms that became multinationals during the interwar period tended to adopt country-centered strategies. European multinationals, operating in a setting where there were many sovereign countries within a relatively small geographical area, were very early to establish self-contained and quite autonomous subsidiaries in many countries. A more tolerant regulatory environment also encouraged European firms to form cartels and other cooperative agreements among themselves, which limited their foreign market entry.

Between the 1950s and the late 1970s, however, there was a strong reversal of the interwar trends. As Figure 2–1 illustrated, very strong underlying forces have driven the globalization of industries. The important reasons can be understood using the configuration/coordination dichotomy. The competitive advantage of competing worldwide from concentrated activities rose sharply, while concentration costs fell. There was a renewed rise in scale economies in many activities due to advancing technology. The minimum efficient scale of an auto assembly plant more than tripled between 1960 and 1975, for example, while the average cost of developing a new drug more than quadrupled. The pace of technological change has increased, creating more incentive to amortize R&D costs against worldwide sales.

Product needs have continued to homogenize among countries as income differences have narrowed, information and communication has flowed more freely around the world, and travel has increased.[11] Growing similarities in business practices and marketing systems (such as chain stores) in different countries have also been a facilitating factor in homogenizing needs. Within countries there has been a parallel trend toward greater market segmentation, which some observers see as contradictory to the view that product needs in different countries are becoming similar. However, segments today seem based less on country differences and more on buyer differences that transcend country boundaries, such as demographic, user industry, or income groups. Many firms successfully employ global focus strategies in which they serve a narrow segment of an industry worldwide, as do Daimler-Benz and Rolex.

Another driver of post–World War II globalization has been a sharp reduction in the real costs of transportation. This has occurred through innovations in transportation technology, including increasingly large bulk carriers, container ships, and larger more efficient aircraft. At the same time, government impediments to global configuration/coordination have been falling in the postwar period. Tariff barriers have gone down, international cartels and patent-sharing agreements have disappeared, and regional economic pacts such as the European Community have emerged to facilitate trade and investment, albeit imperfectly.

The ability to coordinate globally has also risen markedly in the postwar period. Perhaps the most striking reason is falling communication costs, in voice, data, and travel time for individuals. The ability to coordinate activities in different countries has also been facilitated by growing similarities among countries in marketing systems, business practices, and infrastructure—country after country has developed supermarkets and mass distributors, television advertising, and so on. Greater international mobility of buyers and information has raised the payout to coordinating how a firm does business around the world. Increasing numbers of firms who are themselves multinational has created growing possibilities for differentiation by suppliers who were global.

The forces underlying globalization have been self-reinforcing. The globalization of firms' strategies has contributed to the homogenization of buyer needs and business practices. Early global competitors must frequently stimulate the demand for uniform global varieties—for example, as Becton Dickinson did in disposable syringes and Honda did in motorcycles. Similarly globalization of industries begets globalization of supplier industries—the increasing globalization of automotive component suppliers is a good example. Pioneering global competitors also stimulate the development and growth of international telecommunication infrastructure as well as the creation of global advertising media—such as the *Economist* and the *Wall Street Journal*.

Strategic Implications of Globalization

When the pattern of international competition shifts from multidomestic to global, there are many implications for the strategy of international firms. Although a full treatment is beyond the scope of this paper, I will sketch some of the implications here.[12]

At the broadest level, globalization casts new light on many issues that have long been of interest to students of international business. In areas such as international finance, marketing and business-government relations, the emphasis in the literature has been on the unique problems of adapting to local conditions and ways of doing business in a foreign country in a foreign currency. In a global industry these concerns must be supplemented with an overriding focus on the ways and means of international configuration and coordination. In government relations, for example, the focus must shift from stand-alone negotiations with host countries (appropriate in multidomestic competition) to a recognition that negotiations in one country will both affect other countries and be shaped by possibilities for performing activities in other countries. In finance, measuring the performance of subsidiaries must be modified to reflect the contribution of one subsidiary to another's cost position or differentiation in a global strategy, instead of viewing each subsidiary as a stand-alone unit. In battling with global competitors, it may be appropriate in some countries to accept low profits indefinitely—in multidomestic competition this would be unjustified.[13] In global industries, the overall system matters as much or more than the country.

Of the many other implications of globalization for the firm, there are two of such significance that they deserve some treatment here. The first is the role of *coalitions* in global strategy. A coalition is a long-term agreement linking firms but falling short of merger. I use the term *coalition* to encompass a whole variety of arrangements that include joint ventures, licenses, supply agreements, and many other kinds of interfirm relationships. Such interfirm agreements have been receiving more attention in the academic literature, although each form of agreement has been looked at separately and the focus has been largely domestic.[14] International coalitions, linking firms in the same industry band in different countries, have become an important part of international strategy in the past decade.

International coalitions are a way of configuring activities in the value chain on a worldwide basis jointly with a partner. International coalitions are proliferating rapidly and are present in many industries.[15] There is a particularly high incidence in automobiles, aircraft, aircraft engines, robotics, consumer electronics, semiconductors, and pharmaceuticals. Although international coalitions have long been present, their character has been changing. Historically, a firm from a developed country formed a coalition with a firm in a lesser-developed country to perform marketing activities in that country. Today, we observe more and more coalitions in which two firms from developed countries are teaming up to serve the world, as well as coalitions that extend beyond marketing activities to encompass activities throughout the value chain and multiple activities.[16] Production and R&D coalitions are very common, for example.

Coalitions are a natural consequence of globalization and need for an integrated worldwide strategy. The same forces that lead to globalization will prompt the formation of coalitions as firms confront the barriers to establishing a global strategy of their own. The difficulties of gaining access to foreign markets and in surmounting scale and learning thresholds in production, technology development, and other activities have led many firms to team up with others. In many industries, coalitions can be a transitional state in the adjustment of firms to globalization, reflecting the need of firms to catch up in technology, cure short-term imbalances between their global production networks and exchange rates, and accelerate the process of foreign market entry. Many coalitions are likely to persist in some form, however.

There are benefits and costs of coalitions as well as difficult implementation problems in making them succeed that I have discussed elsewhere. How to choose and manage coalitions is among the most interesting questions in international strategy today. When one speaks to managers about coalitions, almost all have tales of disaster that vividly illustrate that coalitions often do not succeed. Also clear is the added burden of coordinating global strategy with a coalition partner because the partner often wants to do things its own way. Yet in the face of copious corporate experience that coalitions do not work and a growing economics literature on transaction costs and contractual failures, we see a proliferation of coalitions today of the most difficult kind—those between companies in different countries.[17] There is a great need for research in both the academic community and in the corporate world about coalitions and how to manage them. They are increasingly being forced on firms today by new competitive circumstances.

A second area where globalization carries particular importance is in organization structure. The need to configure and coordinate globally in complex ways creates some obvious organizational challenges.[18] Any organization structure for competing internationally has to balance two dimensions; there has to be a country dimension because some activities are inherently performed in the country, and there has to be a global dimension because the advantages of global configuration/coordination must be achieved. In a global industry, the ultimate authority must represent the global dimension if a global strategy is to prevail. However, there are tremendous pressures within any international firm once it disperses any activities to disperse more. Moreover, forces are unleashed that lead subsidiaries to seek growing autonomy. Local country managers will have a natural tendency to emphasize how different their country is and their consequent need for local tailoring and control over more activities in the value chain. Country managers will be loathe to give up control to outside forces over activities or how they are performed. They will also frequently paint an ominous picture of host government concerns about local content and requirements for local presence. Corporate incentive systems frequently encourage such behavior, by linking incentives narrowly to subsidiary results.

In successful global competitors, an environment is created in which the local managers seek to exploit similarities across countries rather than emphasize differences. They view the firm's global presence as an advantage to be tapped for their local gain. Adept global competitors often go to great lengths to devise ways of circumventing or adapting to local differences while preserving the advantages of the similarities. A good example is Canon's personal copier (PC). In Japan the typical paper size is bigger than U.S. legal size and the standard European size. Canon's PC copier will not handle this size—a Japanese company introduced a product that did not meet its home market needs in the world's largest market for small copiers! Canon gathered its marketing managers from around the world and cataloged market needs in each country. They found that capacity to copy the large Japanese paper was needed only in Japan. In consultation with design and manufacturing engineers, it was determined that building this feature into the PC would significantly increase its complexity and cost. The decision was made to omit the feature because the price elasticity of demand for the PC was judged to be high. But this was not the end of the deliberations. Canon's management then set out to find a way to make the PC saleable in Japan. The answer that emerged was to add another feature to the copier—the ability to copy business cards—which both added little cost and was particularly valuable in Japan. This case illustrates the principle of looking for the similarities in needs among countries and in finding ways of creating similarities, not emphasizing the differences.

Such a change in orientation is something that typically occurs only grudgingly in a multinational company, particularly if it has historically operated in a country-centered mode as has been the case with early U.S. and European multinationals. Achieving such a reorientation requires first that managers recognize that competitive success demands exploiting the advantages of a global strategy. Regular contact and discussion among subsidiary managers seems to be a prerequisite, as are information systems that allow operations in different countries to be compared.[19] This can be followed by programs for exchanging information and sharing know-how and then more complex forms of coordination. Ultimately, reconfiguring of activities globally may then be accepted, even though subsidiaries may have to give up control over some activities in the process.

The Future of International Competition

Since the late 1970s there have been some gradual but significant changes in the pattern of international competition that carry important implications for international strategy. Our framework provides a template with which we can examine these changes and probe their significance. The factors shaping the global configuration of activities by firms are developing in ways that contrast with the trends of the previous thirty years. Homogenization of product needs among countries appears to be continuing, though segmentation within countries is as well. As a result, consumer packaged goods are becoming increasingly prone toward globalization, though they have long been characterized by multidomestic competition. There are also signs of globalization in some service industries as the introduction of information technology creates scale economies in support activities and facilitates coordination in primary activities. Global service firms are reaping advantages in hardware and software development as well as procurement.

In many industries, however, limits have been reached in the scale economies that have been driving the concentration of activities. These limits grow out of classic diseconomies of scale that arise in very large facilities, as well as new, more flexible technology in manufacturing and other activities that is often not as scale sensitive as previous methods. At the same time, though, flexible manufacturing allows the production of multiple varieties (to serve different countries) in a single plant. This may encourage new movement toward globalization in industries in which product differences among countries have remained significant and have blocked globalization in the past.

There also appear to be some limits to further decline in transport costs, as innovations such as containerization, bulk ships, and larger aircraft have run their course. However, a parallel trend toward smaller, lighter products and components may keep some downward pressure on transport costs. The biggest change in the benefits and costs of concen-trated configuration has been the sharp rise in protectionism in recent years and the resulting rise in nontariff barriers, harkening back to the 1920s. As a group, these factors point to less need and less opportunity for highly concentrated configurations of activities.

When we examine the coordination dimension, the picture looks quite different. Communication and coordination costs are dropping sharply, driven by breathtaking advances in information systems and telecom-munication technology. We have just seen the beginning of developments in this area, which are spreading throughout the value chain (see Porter and Millar 1985). Boeing, for example, is employing computer-aided design technology to jointly design components on-line with foreign suppliers. Engineers in different countries are communicating via com-puter screens. Marketing systems and business practices continue to homogenize, facilitating the coordination of activities in different coun-tries. The mobility of buyers and information is also growing rapidly, greasing the international spread of brand reputations and enhancing the importance of consistency in the way activities are performed worldwide. Increasing numbers of multinational and global firms are begetting globali-zation by their suppliers. There is also a sharp rise in the computerization of manufacturing as well as other activities throughout the value chain, which greatly facilitates coordination among dispersed sites.

The imperative of global strategy is shifting, then, in ways that will require a rebalancing of configuration and coordination. Concentrating activities is less necessary in economic terms and less possible as govern-ments force more dispersion. These forces are pushing firms to inter-mediate positions on the configuration axis as shown in Figure 2–6. At the same time, the ability to coordinate globally throughout the value chain is increasing dramatically through modern technology. The need to coordinate is also rising to offset greater dispersion and to respond to buyer needs. Thus, today's game of global strategy seems increasingly to be a game of coordination—getting more and more dispersed production facilities, R&D laboratories, and marketing activities to truly work to-gether. Yet widespread coordination is the exception rather than the rule today in many multinationals, as I have noted. The imperative for coordination raises many questions for organizational structure and is complicated even more when the firm has built its global system using coalitions with independent firms.

Japan has clearly been the winner in the postwar globalization of competition. Japan's firms had not only an initial labor cost advantage but also the orientation and skills to translate this into more durable competitive advantages such as scale and proprietary technology. The Japanese context also offered an excellent platform for globalization in

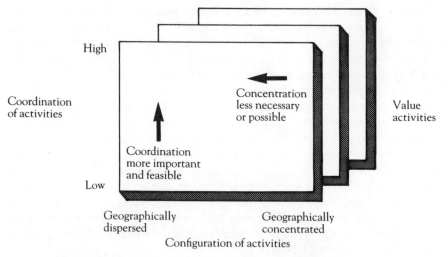

Figure 2–6. Future Trends in International Competition

many industries, given postwar environmental and technological trends. With home market conditions favoring compactness, a lead in coping with high energy costs, and a national conviction to raise quality, Japan has proved a fertile incubator of global leaders. Japanese multinationals had the advantage of embarking on international strategies in the 1950s and 1960s when the imperatives for a global approach to strategy were beginning to accelerate, but without the legacy of past international investments and modes of behavior.[20] Japanese firms also had an orientation toward highly concentrated activities that fit the strategic imperative of the time. Most European and many U.S. multinationals, conversely, were well established internationally before the war. They had legacies of local subsidiary autonomy that reflected the interwar environment. As Japanese firms spread internationally, they dispersed activities only grudgingly and engaged in extensive global coordination. European and country-centered U.S. companies struggled to rationalize overly dispersed configurations of activities and to boost the level of global coordination among foreign units. They found the decentralized organization structures so fashionable in the 1960s and 1970s to be a hindrance to doing so.

As today's international firms contemplate the future, Japanese firms are rapidly dispersing activities, due largely to protectionist pressures but also because of the changing economic factors I have described. They will have to learn the lessons of managing overseas activities that many European and U.S. firms learned long ago. However, Japanese firms enjoy an organizational style that is supportive of coordination and a strong commitment to introducing new technologies such as information systems that facilitate it. European firms must still overcome their country-centered heritage. Many still do not compete with truly global strategies and lack modern technology. Moreover, the large number of coalitions formed by European firms must overcome the barriers to coordination if they are not to prove ultimately limiting. The European advantage may well be in exploiting an acute and well-developed sensitivity to local market conditions as well as a superior ability to work with host governments. By using modern flexible manufacturing technology and computerizing elsewhere in the value chain, European firms may be able to serve global segments and better differentiate products.

Many U.S. firms tend to fall somewhere in between the European and Japanese situations. Their awareness of international competition has risen dramatically in recent years, and efforts at creating global strategies are more widespread. The U.S. challenge is to catch the Japanese in a variety of technologies, as well as to learn how to gain the benefits of coordinating among dispersed units instead of becoming trapped by the myths of decentralization. The changing pattern of international competition is creating an environment in which no competitor can afford to allow country parochialism to impede its ability to turn a worldwide position a competitive edge.

NOTES

1. There are many books on the theory and management of the multinational, which are too numerous to cite here. For an excellent survey of literature, see Caves (1982).
2. Vernon himself, among others, has raised questions about how general the product cycle pattern is today.
3. The distinction between multidomestic and global competition and some of its strategic implications were described in Hout, Porter, and Rudden (1982).
4. Perlmutter's (1969) concept of ethnocentric, polycentric, and geocentric multi-nationals takes the *firm* not the industry as the unit of analysis and is decoupled from industry structure. It focuses on management attitudes, the nationality of executives, and other aspects of organization. Perlmutter presents ethnocentric, polycentric, and geocentric as stages of an organization's development as a multinational, with geocentric as the goal. A later paper (Wind, Douglas, and Perlmutter 1973) tempers this conclusion based on the fact that some companies may not have the required sophistication in marketing to attempt a geocentric strategy. Products embedded in the lifestyle or culture of a country are also identified as less susceptible to geocentrism. The Perlmutter et. al. view does not link management orientation to industry structure and strategy. International strategy should grow out of the net competitive advantage in a global industry of different types of worldwide coordination. In some industries, a country-centered strategy, roughly analogous to Perlmutter's polycentric idea, may be the best strategy irrespective of company size and international experience. Conversely, a global strategy may be imperative given the competitive advantage that accrues from it. Industry and strategy should define the organization approach, not vice versa.

5. Buzzell (1968), Pryor (1965), and Wind, Douglas, and Perlmutter (1973) point out that national differences are in most cases more critical with respect to marketing than with production and finance. This generalization reflects the fact that marketing activities are often inherently country-based. However, this generalization is not reliable because in many industries, production and other activities are widely dispersed.
6. A number of authors have framed the globalization of industries in terms of the balance between imperatives for global integration and imperatives for national responsiveness, a useful distinction (see Prahalad 1975; Doz 1976; and Bartlett 1979). I link the distinction here to where and how a firm performs the activities in the value chain internationally.
7. The difficulties in coordinating are internationally parallel to those in coordinating across business units competing in different industries within the diversified firm (see Porter 1985b: ch. 11).
8. For an interesting description of the industry, see Yoshino (1986).
9. The firm need not necessarily be owned by investors in the country, but the country is its home base for competing in a particular industry.
10. See Chandler (1986) for a penetrating history of the origins of the large industrial firm and its expansion abroad, which is consistent with the discussion here.
11. Theodore Levitt's (1983) article provides a supporting view.
12. The implications of the shift from multidomestic to global competition were the theme of a series of papers on each functional area of the firm prepared for the Harvard Business School Colloquium on Competition in Global Industries (see Porter 1986).
13. For a discussion, see Hout, Porter, and Rudden (1982). For a recent treatment, see Hamel and Prahalad (1985).
14. See Porter and Fuller (1986) for some discussion of organizational alternatives.
15. For a treatment of coalitions from this perspective, see Porter and Fuller (1986).
16. Hladik's (1984) recent study of international joint venture provides supporting evidence.
17. For the seminal work on contractual failures, see Williamson (1975).
18. For a thorough and sophisticated treatment, see Bartlett (1986).
19. For a good discussion of the mechanisms for facilitating international coordination in operations and technology development, see Flaherty (1986). Flaherty stresses the importance of information systems and the many dimensions that valuable coordination can take.
20. What prewar international sales enjoyed by Japanese firms were handled largely through trading companies (see Chandler 1986).

REFERENCES

Bartlett, Christopher A. 1979. "Multinational Structural Evolution: The Changing Decision Environment in the International Division." Unpublished DBA dissertation, Harvard Business School.
———. 1986. "Building and Managing the Transnational: The New Organizational Challenge." In Competition in Global Industries, edited by Michael E. Porter. Boston, Mass.: Harvard Business School Press.
Buzzell, R.D. 1968. "Can You Standardize Multinational Marketing?" Harvard Business Review (November/December): 102–13.
Casson, Mark C. 1982. "Transaction Costs and the Theory of the Multinational Enterprise." In New Theories of the Multinational Enterprise, edited by Alan Rugman. London: Croom Helm.

Caves, R.E. 1982. *Multinational Enterprise and Economic Analysis.* Cambridge: Cambridge University Press.

Chandler, Alfred D., Jr. 1986. "The Evolution of Modern Global Competition." In *Competition and Global Industries*, edited by Michael E. Porter. Boston, Mass.: Harvard Business School Press.

Doz, Yves. 1976. "National Policies and Multinational Management." Unpublished DBA dissertation, Harvard Business School.

Flaherty, M.T. 1986. "Coordinating International Manufacturing and Technology." In *Competition in Global Industries*, edited by Michael E. Porter. Boston, Mass.: Harvard Business School Press.

Hamel, G., and C.K. Prahalad. 1985. "Do You Really Have a Global Strategy?" *Harvard Business Review* (July/August): 139–48.

Hladik, K. 1984. "International Joint Ventures: An Empirical Investigation into the Characteristics of Recent U.S.-Foreign Joint Venture Partnerships." Unpublished doctoral dissertation, Business Economics Program.

Hout, T., M.E. Porter, and E. Rudden. 1982. "How Global Companies Win Out." *Harvard Business Review* (September/October): 98–108.

Levitt, Theodore. 1982. "The Globalization of Markets." *Harvard Business Review* (May/June): 92–108.

Perlmutter, Howard V. 1969. "The Tortuous Evolution of the Multinational Corporation." *Columbia Journal of World Business* (January/February): 9–18.

Porter, Michael E. 1980. *Competitive Strategy: Techniques for Analyzing Industries and Competitors.* New York: Free Press.

———. 1985a. "Beyond Comparative Advantage." Working paper, Harvard Business School.

———. 1985b. *Competitive Advantage: Creating and Sustaining Superior Performance.* New York: Free Press.

———, ed. 1986. *Competition in Global Industries.* Boston, Mass.: Harvard Business School Press.

Porter, Michael E., and Mark B. Fuller. 1986. "Coalitions and Global Strategy." In *Competition in Global Industries*, edited by Michael E. Porter. Boston, Mass.: Harvard Business School Press.

Porter, Michael E., and Victor Millar. 1985. "How Information Gives You Competitive Advantage." *Harvard Business Review* (July/August): 149–60.

Prahalad, C.K. 1975. "The Strategic Process in a Multinational Corporation." Unpublished DBA dissertation, Harvard Business School.

Pryor, Millard H. 1965. "Planning in a World-Wide Business." *Harvard Business Review* 43 (January/February): 130–39.

Teece, David J. 1986. "Transactions Cost Economics and the Multinational Enterprise." *Journal of Economic Behavior and Organization* 7: 21–45.

United Nations. Various years. *Statistical Yearbooks.* New York: United Nations.

United Nations Center on Transnational Corporations. 1984. *Salient Features and Trends in Foreign Direct Investment.* New York: United Nations.

Vernon, Raymond. 1966. "International Investment and International Trade in the Product Cycle." *Quarterly Journal of Economics* 80 (May): 190–207.

Williamson, Oliver. 1975. *Markets and Hierarchies.* New York: Free Press.

Wind, Yoram, Susan P. Douglas, and Howard B. Perlmutter. 1973. "Guidelines for Developing International Marketing Strategies." *Journal of Marketing* 37 (April): 14–23.

Yoshino, Michael. 1986. "Global Competition in a Salient Industry: The Case of Civil Aircraft." In *Competition in Global Industries*, edited by Michael E. Porter. Boston, Mass.: Harvard Business School Press.

3 INVESTMENT IN NEW TECHNOLOGY AND COMPETITIVE ADVANTAGE

Kim Clark

The decision to invest in new technology—to acquire and apply new technical knowledge and capabilities—is among the most important competitive decisions that managements must make. Investment in innovation has always been a significant aspect of competition in emerging science-based industries like artificial intelligence, bioengineering, and advanced materials. But I believe that changes underway in technology and international competition will enhance the importance of both innovation in established and emerging manufacturing industries and sharpen the focus on its competitive role. What has happened in manufacturing industries is the emergence of a new kind of competition. Where the competitive arena was once the domestic market, and the competitors were once familiar domestic firms, today new international competitors with advantages in cost or productivity or government support have emerged. At the same time, new concepts in products and processes have altered the value of existing commitments and created new possibilities in marketing and production. The thrust of these changes has been intense competitive pressure on the established firms whose roots are in the old competitive arena and the old technologies. Managerial concepts, relationships with suppliers and with employees, technical skills, and equipment and processes that may have once served well are no longer viable.

For established manufacturers, the challenge of the new competitive environment is evident in greater pressure on margins, shorter product life cycles, and new standards of performance and quality. In the midst of such difficulty it is hard to see the opportunities inherent in the new competitive environment. But the forces driving change—new demands in the market place, new technologies—create opportunities as well as challenges. There may well be some reduction in size as old assets are replaced, and some firms may not be able to make thorough changes in concepts and practices. But those that can adapt their strategic vision to the new competitive realities and develop new capabilities may emerge more vigorous, more competitive than before.

At the heart of a vigorous and successful response to the challenge of the new industrial competition—whether in steel, autos, semiconductors, or telecommunications—is a commitment to new concepts in managing production, new approaches to customers and markets, and the application of new technology. All these innovations entail investment of a special kind. For what is critical in the new industrial competition is to invest in new technology that creates (or has the potential to create) competitive advantage.

Investment projects are typically reviewed through some kind of financial framework to determine their effect on revenues, costs, assets, and profitability. But the question of competitive advantage requires additional consideration. In this chapter I develop a way of thinking about the competitive role of investment in new technology. I use examples from the steel industry to illustrate the kinds of issues that arise in evaluating competitive significance, but the concepts are general. The central theme is that there are different kinds of innovation, with different requirements for success; the implication is that different kinds of innovation require different kinds of analysis.

The first part of the chapter lays out the conceptual framework, and introduces the notion of *transilience*—an innovation's capacity to transform existing systems of technology and marketing. Application of the framework generates four different types of innovation—architectural, regular, niche creation, revolutionary—that are illustrated with examples from the steel industry. The steel industry innovations suggest both the distinctions between the types of innovation and the considerations appropriate in decisionmaking about each. The third part of the chapter focuses on the salient differences in the four modes of innovation. The perspective is that of a firm faced with the questions of whether to introduce the technology and how to do it. The chapter concludes with brief observations on the implications of the analysis for the organization and conduct of decisionmaking about new technology and suggestions for subsequent research.

Technology and Competitive Advantage

The impact of a change in technology on the competitive advantage of the innovating firm depends on how the innovation affects the firm's products and services. Firms compete by offering products that may differ in several respects: performance, reliability, availability, ease of use, aesthetic appearance, and, of course, cost. A firm gains a competitive advantage when it achieves a position in one of these featured dimensions, or a combination of them, that is both valued by customers and superior to that of its competitors. It is important to note, however, that the product features themselves are not the fundamental source of advantage. The foundation of a firm's competitive position is the set of capabilities (material resources, human skills and relationships, and relevant knowledge) the firm uses to build the product features that appeal to the marketplace. The significance of a change in technology for competitive advantage thus depends on what I shall call its transilience—that is, its capacity to influence the firm's existing resources, skills, and knowledge.

Table 3-1 presents a list of the principal aspects of the firm's competitive capability. In the top half of the table I have placed the factors that determine the capabilities of the firm in technology and production; I shall refer to this as the technology domain but mean it to include production and operations. The resources, skills, and knowledge within this domain are linked to competition through their effect on the physical characteristics of the product—its performance, appearance, quality, and so on—and its cost. The list includes traditional factors of production like materials, people, and equipment, as well as knowledge and experience relevant to design and production. This not only includes links to scientific, engineering, and design disciplines, but it also includes the knowledge embedded in the systems and procedures used to organize production.

The second half of Table 3-1 focuses on linkages to markets and customers. The definition of the customer group and the strength of the firm's relationship with the customer determine the size and character of the market. They not only influence the overall size of potential sales but the firm's ability to understand customers and meet customer expectations. Other aspects of the customer/market linkage involve how the customer acquires and uses the product and the nature of the services the product provides. As in the technology domain, these aspects of the market domain influence customer choice (and thus the firm's relative position) through product characteristics like performance (application, knowledge), availability (channels of distribution), and ease of use (communication, knowledge, service).

Each item listed in the table is accompanied by a scale that depicts the range of effects an innovation might have. The range is defined by polar extremes, the one conservative, the other radical. On the conservative end of the scale are innovations that enhance the value or applicability of the firm's existing capability. Clearly, all technological innovation imposes change of some kind, but change need not be destructive. Inno-

Table 3–1.   Innovation and Firm Competence

Domain of Innovative Activity		Range of Impact of Innovation
I. *Technology/production*		
Design/embodiment of technology	Improves/perfects established design	<-> Offers new design/radical departure from past embodiment
Production systems/organization	Strengthens existing structure	<-> Makes existing structure obsolete; demands new system, procedures, organization
Skills (labor, managerial, technical)	Extends viability of existing skills	<-> Destroys value of existing expertise
Materials/supplier relations	Reinforces application of current materials/suppliers	<-> Extensive material substitution; opening new relations with new vendors
Capital equipment	Extends existing capital	<-> Extensive replacement of existing capital with new types of equipment
Knowledge and experience base	Builds on and reinforces applicability of existing knowledge	<-> Establishes links to whole new scientific discipline/ destroys value of existing knowledge base
II. *Market/customer*		
Relationship with customer base	Strengthens ties with established group	<-> Attracts extensive new customer group/creates new market
Customer applications	Improves service in established application	<-> Creates new set of applications/new set of customer needs
Channels of distribution and service	Builds on and enhances the effectiveness of established distribution network/service organization	<-> Requires new channels of distribution/new service, after market support
Customer knowledge	Uses and extends customer knowledge and experience in established product	<-> Intensive new knowledge demand of customer; destroys value of customer experience
Modes of customer communication	Reinforces existing modes/methods of communication	<-> Totally new modes of communication required (such as, field sales engineers)

vation in technology may solve problems or eliminate flaws in a design or process that makes existing resources or skills more effective. Such change conserves the established competence of the firm, and if the enhancement or refinement is considerable, may actually entrench those skills making it more difficult for alternative resources or skills to achieve an advantage.

On the radical end of the scale the effect of innovation on established capabilities is quite the opposite. Instead of enhancing and strengthening, innovation of this sort disrupts and destroys. It changes the technology of process or product in a way that imposes requirements that the existing resources satisfy poorly or not at all. The effect is thus to reduce the value of existing competence and in the extreme case to render it obsolete. Its effect on competition works through a redefinition of what is required to achieve competitive advantage.

In both the market and technology domains the effect of innovation on competition depends on what it does to the value and applicability of established capabilities—that is, on its transilience. Competitive effects are not limited to the kind of radical innovations that create new industries. By classifying an innovation as conservative, I do not mean that it is marginal, timid, or trivial but that it builds on established skills and resources. It also may mean that the firm's position in the market is entrenched, so that potential and actual competitors are placed at a disadvantage. Depending on what resources and skills competitors have, a conservative innovation could have significant competitive consequences.

It is likewise important to maintain the distinction between market and technology domains in assessing competitive effect. A given innovation may affect the two kinds of capability quite differently. A new alloy, for example, might require refinement of existing processes in the technology domain but create new customers and new applications in the market domain. The important thing to note is that it is the pattern of effects in the two domains that determines competitive impact. One way to depict the pattern of effects is to use composite transilience scales for each domain as the axes of a two-dimensional diagram. In Figure 3–1 I have positioned the market transilience scale in the vertical dimension and the technology transilience scale in the horizontal. This creates a transilience map with four quadrants representing a different type of innovation.

Working counterclockwise from the upper right-hand corner, the categories of innovation are as follows:

Architectural: radical technology applied to new markets;
Niche creation: refinements in technology applied to new customer groups and new applications;

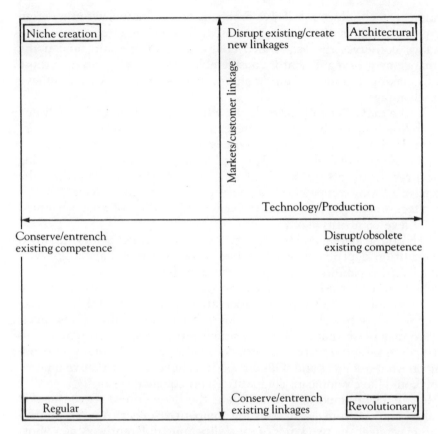

Figure 3–1.   Transilience Map.

Regular: refinements in technology applied to existing markets and customers;

Revolutionary: disruptive change in technology applied to existing markets and customers.

There is more to the transilience map than a simple categorization of innovation. The different kinds of innovation are also closely related to different kinds of competition and competitive strategy and depend as well on different kinds of managerial and organizational support. Moreover, the four types of innovation differ in the way that they create both value and uncertainty and in the supporting investments they require. All these differences have important implications for the analysis of investment in new technology.

The Transilience of Innovation:
Examples from the Steel Industry

This section of the chapter examines four innovations in steel production. The examples illustrate the differences among the quadrants of the transilience map and highlight the links between innovation and competition. Each of the innovations is treated as a project, and the decision confronting the prospective innovator is examined using the transilience framework. I have provided a minimum of technical background to put the innovation in perspective while focusing the bulk of the discussion on the implications for competition and management.

Architectural Innovation: The Bessemer Process

Prior to 1860 steel was produced in very limited quantities using batch processes that were time consuming and expensive. In the single crucible process wrought iron and charcoal were placed in an airtight crucible and brought to a molten state. Although this process was a great improvement over the cementation process (that is, baking solid bars of iron in charcoal), which often required a week to complete, the crucible process was not capable of high-volume production. As a result steel was limited to special applications prior to 1860. All this changed with the development of the Bessemer process for steelmaking in the late 1850s. In this process air was blown through a bath of molten iron creating a combustible mixture of oxygen, carbon, silicon, and manganese. Complete combustion left behind iron with very little carbon, silicon, and manganese, but small amounts of carbon and manganese were added back to make steel.

A project to innovate in Bessemer steelmaking in 1870 involved construction of special buildings and installation of a Bessemer converter with associated ladles, casting pit, molds, hydraulic machinery, and so forth. In 1870 the cost of the initial investment in plant and equipment would have been about $200,000. The innovating firm was likely to have experience in iron production and contacts with iron and (potential) steel customers.

The Bessemer process was an architectural innovation. It marked a radical departure from the small-scale batch processes of the day, and by breaking the constraints on speed and volume it laid the foundation for new markets and applications. There were few aspects of the production system that were unaffected. The new process introduced short-cycle production (twenty minutes versus many hours) that required new equipment and new procedures. Timing of actions and precision in execution became critical requirements for production labor and management. Control of the chemical properties of the material became significant but the process was more difficult to control. The greater speed of production and the vastly enlarged scale of output per heat (2 to 2.5 tons every twenty minutes) had important ramifications on upstream and downstream processes.

On the market side, the effects were perhaps even more extensive (though not more dramatic—the Bessemer process was a pyrotechnical extravaganza). With several tons of steel coming out of a converter every few minutes, the new process made available large quantities of steel at relatively low cost. The innovation thus opened up new markets, applications, and customers. Even where steel was used as a straightforward substitute for iron, its superior properties expanded the size and scope of markets. In the market for nails, for example, Bessemer steel wire supplanted iron plate as the basic raw material because of its greater strength and ease of use. And there were many completely new applications. In construction, for example, Bessemer steel was an important element in the development of the skyscraper: The strength and ductility of steel made possible skeleton construction methods using rivets for joining.

In this and a variety of other uses, the Bessemer process opened up new markets, created the need for new channels of distribution, brought iron- and steelmakers into contact with new customers and new demands for performance and reliability. In the sense of the transilience map, and in the sense of its role in laying the foundation for the modern steel industry, the Bessemer process was an architectural innovation.

It would be a mistake, however, to treat the Bessemer process as a single innovation. To understand the nature of architectural innovations and to understand the decisions facing firms and entrepreneurs who tried to develop and exploit the innovation, one must see it as a cluster of innovations that are introduced over a period of time. Moreover, the innovations involved are partly technical, partly organizational, partly procedural. The theme that binds all of them together is the creation of new capabilities within the firm that allows the firm to unlock or exploit the potential in the technology.

New capabilities were essential to the success of the Bessemer process. When first introduced the process was unstable, dangerous, and fraught with uncertainty. Poor refractory material, inadequate design, inappropriate procedures, and lack of the right equipment made early Bessemer converters prone to explosions, spills of molten steel, and almost constant repair, adjustment, and cleanup. It was not until new concepts in plant design, equipment, and procedure were developed that steelmakers were able to harness the power of the technology.

The basic idea behind the process was clear enough at the very beginning, but steelmakers faced a variety of technical and procedural options in making the process a commercial reality. Thus, the early period of Bessemer production (from 1864 to 1880) was a time of experimentation and search. On the technical side, engineers introduced new equipment like elevated converters and removable bottoms; they rearranged plant layouts to give more continuity of operation; and they introduced new procedures in the timing of maintenance and the sequence of melting and pouring. In addition, firms set up laboratories and began to study the chemistry of steel production and began to apply their new understanding to the mix of ingredients used in the process. Under the relentless drive for new insight and understanding, explosions and accidents became rare events, the process became more predictable and controllable, output expanded dramatically, and productivity of both capital and labor grew rapidly. These developments were found in both Europe and the United States, but it was the U.S. firms and their close attention to innovation in equipment and operating practice that pushed the process furthest: In 1901 eighty-one converters in the United States produced 9.8 million tons of steel, while seventy-eight converters in the United Kingdom produced less than 2.0 million tons.

The pattern of innovation in technology and practice that created the Bessemer steel industry was the fruit of organizational and managerial capabilities. Those individuals and organizations that made the decision to invest in the Bessemer technology were in fact deciding to develop a distinctive set of capabilities. The pursuit of and application of better understanding of the process required the ability to solve problems involving equipment and procedure. And the problems could not be identified or solved apart from actual production. This was problem solving in "real time." Moreover, it involved experimentation with production equipment. Firms had to learn and produce at the same time. Those that succeeded combined engineers with hands-on experience in production and steelmakers with a commitment to doing things differently.

Further, the organizations had to develop the ability to change. The shop floor of a Bessemer steel plant in the 1870s was not a tranquil place: It was a time of new equipment, new procedures, and new layouts. Assignments could change, tasks could be redefined, and relationships altered with relative frequency. But there was more. The hallmark of the successful new steel firms was their commitment to pushing out the frontier of technology. It was not just a certain flexibility of mind and organization but a willingness to take on technical (and personal) risks and a vigilance in staying on top of the latest developments in equipment and practice that made the difference.

Regular Innovation: Modern Continuous Casting

Continuous casting is a process that transforms liquid steel into semi-finished slabs, billets, or blooms. It eliminates the need to make ingots and then process the ingots through a slabbing mill. When first introduced in the 1950s, continuous casting was limited in the range of steels that it could process, and there were significant technical problems with yields, surface quality, and continuity of operation. Through a long period of subsequent innovation these problems were solved. The decision to introduce a modern continuous caster thus involves far less technical uncertainty, but there are innovative changes in the equipment that need to be evaluated. In terms of its impact on the technical system and linkages with customers, technical developments in modern continuous casting are the fruits of regular innovation.

Take, for example, the issues facing a firm trying to understand the implications of introducing a new slab caster with rated capacity of 150,000 tons per month. In this example the firm already has continuous casting operations, and the particular equipment under consideration will be designed and built by an outside vendor. The basic concepts in the technology are well known; the vendor has built and installed many similar processes. There are, however, innovations in equipment design and operation. The caster will use a new cooling system for the molds that affords greater control and efficiency; there will be a new variable width mold device that will allow more rapid changeover to different slab sizes; there will be a new approach to protecting the steel from contact with oxygen. All of these innovations are designed to improve quality and efficiency of existing products.

The innovations in the new slab caster extend and refine existing concepts. They build on and make use of existing engineering skills and require very little change in casting practices or procedures. Indeed, they enhance the effectiveness of the firm's established technical and operating resources. The result of higher quality and more efficient operation will be a strengthening of the firm's relationship to its existing customers and a reinforcement of its approach to distribution and market communication. Thus, on both the technical and marketing sides of the business, the impact of the innovation in continuous casting is to conserve and perhaps to entrench existing resources, skills, and relationships.

This does not imply that it is business as usual at the firm that puts in the slab caster, nor that its introduction will not require the application of significant capabilities. The reduced changeover times may make possible changes in scheduling practice and even in the firm's approach to serving customers. Moreover, enhanced quality of semifinished product may have an influence on the approach taken in finishing. The casting process itself may require different kinds of maintenance activities and may pose a different set of equipment control problems, given the new spraying systems, mold technology, and so forth. The important point to note is that these changes are conservative—they make use of and strengthen the firm's established capabilities.

Successful implementation of the technology will require application of skills in vendor management, process engineering, and problem solving. Systems like the slab caster are often built to the specifications of the customer. The effectiveness of the process will thus depend on the firm's ability to relate its operating and commercial requirements to the specifics of process design. There may be further opportunities for improved effectiveness during the installation phase as particular issues in construction and implementation are addressed. Even where the vendor delivers a turnkey system, it is difficult for the vendor to anticipate all of the particular problems the system will encounter or the particular demands that the firm will make of the new process. In fact, it is unlikely that the firm itself will be able to do so. The ability to make changes to the process and the equipment will thus be essential to full utilization of its capabilities.

The same logic applies to improving yields, quality, and efficiency. The capability within the organization to identify problems, develop solutions, and put them into practice will be essential to the ongoing development of the process. Given the nature of the technology and the context in which it is introduced it is likely that our hypothetical firm will have made investments in these kinds of engineering and process development skills. If it has not, or if the existing skills are not adequate, the investment in the equipment and the process will have to be supplemented with investment in training or in new engineering talent.

Revolutionary Innovation: The Continuous Micromill

Since the introduction of the Bessemer process in the 1860s increased continuity of operation has been a central theme in steel process innovation. Compared to the mills of the early 1900s the modern steel mill is far more integrated and continuous, with fewer interruptions of the flow of material and less work-in-process inventory. But the modern mill still has many of the characteristics of a batch process: intermittent flows, work-in-process storage, lots of starting and stopping of processes. A fully continuous process would mark a significant departure from established practice.

One approach to continuous processing is contained in the proposed project to introduce a continuous micromill. As the name implies the project introduces a very small-scale steel mill that operates continuously. The new mill embodies a variety of innovations in technology and in process control and design. Using scrap as a raw material, the melting operation would use a small electric induction furnace that would feed a continuous horizontal caster. The cast billets would be processed directly into the finishing process for the shaping and treating of a variety of products. Although different pieces of this process have been operated on an experimental basis, the system would be an innovation. Several uncertainties surround the design of the furnace, the extent of and type of refining done between the furnace and the caster, the degree of product flexibility in the caster, and the types of finishing operations to be integrated into the process. In addition, there is likely to be a need for an innovative approach to process control.

The mill would be designed to produce something like 50,000 tons per year and would serve a very small local market. The plant would be highly automated and would employ no more than fifty people in all aspects of the business. Because of the technology involved and the degree of automation, the workforce is likely to be largely made up of engineers and skilled technicians. The organization is likely to involve very little hierarchy, with most activities taking place in work teams. At least in its initial stages the mill would be set up to produce a range of existing products and serve existing markets.

As outlined here the continuous micromill is a revolutionary innovation. As a system, it is a radical departure from existing production processes and would require the development of new knowledge, new equipment, new techniques, and new methods of process control and management. Its potential advantage lies in low capital cost per ton of output, in low labor cost, and in its close proximity to the customer. If successful, it could render some established processes for some products obsolete. Although its initial application is in established markets and products (which places it in the revolutionary category), successful development of the technology could result in the establishment of new channels of distribution and new forms of business relationship. Some firms may find it advantageous to acquire their own micromill to serve their steel needs, or they may find it useful to have steel suppliers follow the example of the can manufacturers and locate close at hand but operate on a contract basis.

Of course, all of this speculation hinges on the viability of the technology and the system. Entry into this technology is fraught with uncertainty. There are a variety of technical uncertainties in basic concept, in the specifics of design, and in the interaction of the different elements in the system. Further, the necessary elements for process control may not have been fully developed, and the particular labor and engineering and management skills, not to mention organization forms, remain to be determined. As with many large-scale innovations only a few of these technical uncertainties can be resolved at the laboratory level or in experimental units. One may have to build a commercial scale mill and develop the technology through experimentation in production.

There are risks as well on the market side. Because the technology will be applied to existing markets, the products of the micromill must meet established commercial standards of cost and quality. The new mill will not have the advantage of offering a wholly new kind of product for which there are few substitutes. It must succeed by being better than the established producers—by offering superior performance, higher reliability, lower cost, or faster delivery. And the firm that undertakes the project must expect established competitors to protect their established positions.

It is clear from this brief discussion that successfully developing the continuous micromill will require specific skills and capabilities in technology and in marketing. On the technical side, the key word is *development*. Some of the equipment may be supplied by vendors, but it seems essential that the innovating firm have the capability to build, investigate, modify, and improve the equipment. In addition, because of the systemic nature of this technology, the firm must be skilled in process control and thus in the management of computer and software vendors and probably in the design and development of software itself. In both systems and equipment, there is likely to be a good deal of experimentation required and the need to apply experimental results to practice. Skill in experimental design and in analysis of experimental data will be essential.

From the standpoint of markets and customers, the decision to undertake this project is a decision to launch a competitive assault on established competitors. The central problem is to discover competitors' weaknesses and target technological and commercial development appropriately. The new technology may provide a basis for changing the rules by which the established competitors play the competitive game. The needs of customers may not be met under existing approaches or may be met more effectively with the aid of the new technical system. Harnessing the potential in the technology in exploiting weak points in established competitors is the central managerial problem. In this context, close collaboration between technologists and marketers is essential. Development of the technology must be guided by the commercial realities of cost, delivery, and quality standards and by the firm's strategy for entry. Yet good technical insight can help to break through existing conventions, suggest new possibilities in product characteristics and customer service, and thus create strategic possibilities.

Niche Creation: The Continuous Heat-Treating Process (CHTP)

The introduction of continuous heat treating of sheet steel is an example of a project that uses refinements and extensions of existing technology to create new market opportunities. In a traditional batch heat-treating process several coils of sheet steel are stacked in a furnace and heated for many hours (a few days in some cases). The continuous process allows for much greater control over the effects of heating and cooling on the microstructure of the steel. The project involves installation of new equipment, automated material flow, and extensive computerized process control. Instead of placing coils in a furnace, the CHTP uncoils the sheet and runs it through a long sequence of cleaning, heating, and cooling operations at relatively high speed; six minutes are required to complete the process for a coil of sheet. The innovation in the system involves both new equipment, new refractory materials, new sequencing, and the use of a complex computer-based adaptive control system. Not only does the CHTP give greater control over microstructure, but it ensures much greater uniformity of structure throughout the coil. In addition, the process is much faster and affords some direct and indirect cost savings as well.

From the standpoint of the existing integrated steel mill, the CHTP builds on and extends established skills, knowledge, and experience. It does require new equipment and a new approach to computer process control, but the bulk of the other steelmaking and steelfinishing processes are not made obsolete. Indeed, the effect on the balance of the process would be to strengthen and conserve its viability. Further, the CHTP's application of computer controls follows a long-term pattern of development in integrated mills. The fact that the entire process can be purchased from outside vendors and linked to existing rolling mills is further evidence that CHTP represents an extension and refinement of existing concepts.

The impact of CHTP on markets and customers depends on how it is used. The new process can be applied to existing customers and markets, but the greater precision and control could turn CHTP into a tool for new product development. By altering the timing and sequence of operations the CHTP can be used to create dual-phase steels and superhigh strength steels that are both ductile and yet very strong. Products with combinations of performance characteristics not currently available in the market could be used to establish new customer linkages and perhaps new market niches.

The creation of new market niches based on CHTP requires capabilities and competence in product development and customer service. Turning the new technology into an effective marketing tool demands an ability to read markets, anticipate customer needs, and direct resources to meeting those needs quickly. The idea is to uncover a need or new application and then move to exploit it quickly before competitors can enter and destroy profitability.

Three kinds of capability seem essential. The first is knowledge and experience in potential applications and understanding of the customer's needs. This is not only required in the existing customer base but among potential customers as well. Second is applications engineering: developing technical solutions to an identified need. This will involve engineering development in the basic product—coming up with a particular composition—and process—designing the process to achieve the design. But it may also involve work in the field of application like stamping operations. Firms attempting to create new niches must work closely with the customer's engineering group and may find it advantageous to help engineer the customer's product or process as well. Finally, niche creation requires a manufacturing operation that is responsive to customer needs and market developments. Compared to the production of long runs of commodity-grade products, niche creation may require shorter runs, shorter lead times, and an ability to accommodate a shifting product mix. Moreover, manufacturing and engineering must also continue to develop the process, refining and strengthening its technical capabilities in order to support product development.

Four Modes of Innovation: Managerial and Competitive Differences

Although investment in new technology is often studied and discussed as though all innovations were cut from the same cloth, the steel cases suggest differences in the kinds of innovation that extend beyond the purely technical. And it is even more than the fact that some innovations entail major change while others are minor. From the standpoint of the firm faced with decisions about innovation, the critical differences are those that affect how the innovation creates both value and risk, how the innovation enters into competition, and how the innovation must be managed. Table 3–2 summarizes the salient differences among the four kinds of innovation, using the four steel projects as examples.

It should be no surprise that several of the critical differences in these investment projects concern the nature of the future. The degree to which future developments in the particular technology itself, in its main alternatives, and in the markets are predictable influences the nature and extent of the uncertainty in a project. The transilience map suggests a natural distinction between technical and market uncertainties, but it also implies differences in the extent to which the process that will generate future outcomes is understood. In both the technology and

Table 3–2. Characteristics of Four Modes of Innovation

Characteristic	Architectural Mode	Regular Mode	Niche-Creation Mode	Revolutionary Mode
Nature of uncertainty/risk	High level of uncertainty about technical options, customer choice, and market definition; Significant uncertainty about competitors; Evolutionary path unclear	Low technical and market uncertainty; Some implementation risk; Evolutionary path relatively clear and well defined	Low technical uncertainty; significant uncertainty about customer choice/market definition; Some implementation risk; business risk because of possible quick response; Evolutionary path unclear on market side	High level of technical uncertainty; low market uncertainty; Implementation risk; significant business risk because of reaction by competitors; Evolutionary path unclear on technical side
Source of value present/future				
Short-term	New conception of product and unprecedented levels and types of performance	Improvement in existing dimensions of competition (cost, quality)	Serving new market segment based on improvement/nuance in performance	Radical improvement in existing product (such as change in price/performance tradeoff)
Long-term	Potential for new industry and subsequent growth and development	Basis for internal improvement along established lines	Product development and further segmentation	New technical evolution; redefine industry
Supporting investments				
Technical	Equipment development; procedure and organization	Training; maintenance system changes; subsequent development	Applications engineering; nimble manufacturing	Equipment development; knowledge of organization; procedures
Market	Information; new channels; relationships; salesforce	New selling points; improved service	Information; new channels; customer service	Information; new selling points

Competitive context/impact				
Basis of competition	Fundamental innovation in product performance	Cost, delivery, quality	Product features and performance, quality	Superior product characteristics
Competitive role	Forge new market/technical configuration; may be basis for entry	Defensive: avoid disadvantage; offensive: consolidate position	Establish temporary monopoly within new customer segment	Basis of assault on established competition—rewrite rules
Competitive advantage	Development of proprietary technology tailored to emerging demands	Implementation; nuances in engineering and sales	Skill in marketing—verify need and tailor technology to fit	Development of technical expertise and skill in application
Competitor reaction	Introduction of alternative	Introduction of similar technology: reinforce positions of comparable firms	Relative quick imitation	Vigorous response by competitors—possible major improvements in established technology approach
Managerial skills	Real time experimentation/search for new ideas Assimilate new ideas/adapt to change Synthesize diverse technologies to meet user need Entrepreneurship in market development	Careful planning/analysis Vendor relationships Directed problem solving within known constraints Equipment adjustment/development	Ability to "read" markets/spot emerging customer demands Rapid product development Nimble manufacturing Customer-driven technology refinement	Leadership in breaking industry conventions Application of scientific and engineering insight Push technology disciplined by goals defined by commercial realities Flexibility in technical and production organization

market domains, conservative innovations involve known alternatives to which probabilities can be assigned. In the continuous casting case, for example, there is so much experience with the technology that the firm can predict the effect on yields and costs with a high degree of confidence.

The same is sometimes true for the more radical changes in technology, but here there is often much less information about technical and market choices. Alternatives may be so ill defined that assigning probabilities would make little sense. Differences in the clarity of alternatives and in understanding of future developments affect not only the overall level of risk in a project, but they also influence where the firm ought to search for information, what kinds of problems need priority in development, and even what kinds of analysis one ought to do in making decisions about the innovation.

If one finds big differences in risk and uncertainty across the quadrants of the transilience map, what of value? Here too the future looms large. All of the innovations create value by improving on existing practice, but they also change the opportunities the firm faces both now and in the future. Moreover, present and future value and opportunity come in different forms and in different combinations in the four modes of innovation. For regular innovation, value is more in immediate improvement in established criteria of performance, whereas the value of a revolutionary innovation may lie far more in the opportunity to redefine the business and exploit future market and technical developments as the new business unfolds.

These future opportunities are distinct from the (relatively certain) cash flows that one can ascribe to the current project. They are possibilities that will depend on other developments and will entail other investments. What is important about them is that without investment in the current project, the future opportunities may not exist. Take, for example, the CHTP project. Development of that capability may lead to the discovery of a new type of steel and create the opportunity for a whole new line of business. The capabilities in the CHTP thus create the opportunity and will be essential in its exploitation.

The value of a given technology depends on both present and future opportunities. Like the assessment of risk, identifying the value of an investment in new technology is complicated by the presence of both technical and market uncertainty. But there is a further issue that confronts both. Risk and value depend not only on what happens in the laboratory or in the minds of the customer but on the actions of competitors—and not only on what competitors do but how the innovating firm uses the innovation and how it deals with competitive response. In this respect, different innovations pose very different challenges and have very different implications for competition, and for management action.

As Table 3–2 makes clear, each of the quadrants of the transilience map is associated with a quite different competitive environment, and each of these kinds of innovation carries with it very different competitive implications. Where architectural innovation, for example, can be (and is often) used by new firms to lay the foundation of a new industry and make the old firms in related businesses obsolete, regular innovation is likely to be an instrument of consolidation, a way for established firms to reinforce their advantages in scale, product performance, and customer service. Moreover, whereas architectural and revolutionary innovation may be used offensively, regular innovation is likely to be a defensive move designed to avoid disadvantage. Further, since regular innovation builds on established practice, it is likely that competitors will be able to imitate the innovation more easily than competitors confronted with a revolutionary change. This implies that whatever competitive advantage a regular innovation may afford will have to be found in the nuances of implementation and subsequent development in both marketing and engineering.

The differences in competitive environment and role have implications for the competitive skills that will be essential to successful innovation. This is clear, for example, in the comparison of niche-creation and revolutionary innovation. The key to creating niches through innovation is to tailor technical development to emerging or perceived customer needs. It is not, as it is in revolutionary innovation, to pursue a new technical agenda. Where the one emphasizes timing and quick response for temporary monopoly, the other focuses on technical breakthroughs that will redefine an industry.

A central implication of competitive differences, and of differences in risk and the creation of value, is that different innovations have to be managed differently. This is obvious if one refers to the development of technology. Clearly, the kind of engineering and of scientific work that will be important in regular innovation will be different than that required in the architectural mode. Both may be handled with great skill, but the former will involve careful planning and analysis and quite directed problem solving, while the latter will involve more experimentation, more trial and error, and more cross-discipline problemsolving. But the differences in management will extend to other activities within the firm and even to its general management. Within each of the traditional functional areas, like manufacturing and marketing, successful innovation in each quadrant requires different approaches and different strengths.

In the case of niche creation, for example, decisions about personnel, incentives, organization, communication, and so forth must reinforce the emphasis on close customer relationships, development of new products, and nimble, responsive production. Furthermore, decisions within each functional area must be consistent with one another. Success in niche-creation innovation will be limited if marketing is pushing for quick delivery to a new customer, while manufacturing is scheduling the plant for long runs to lower cost. What is required for success differs across the quadrants, but within a quadrant successful innovation requires coherence in supporting activities.

It should be evident from this brief comparison of the types of innovation that the introduction of new technology involves much more than the installation of a piece of equipment or a new process. Each of the kinds of innovation entails a whole set of supporting investments, capabilities, and conditions. The implication is that the four quadrants of the transilience map not only depict types of innovation but types of management and organization and modes of competition. Further, those conditions and capabilities are long lived. The transilience framework thus underscores the long-term, developmental nature of innovation decisions. It suggests that a given project to introduce new technology should be seen as a sequence of investments that will take the firm down a particular strategic path.

Seen in these terms, the decision to invest in new technology raises several questions about strategy and competition over the long term. Once it is clear what effect the investment will have on established competence in production, technology, and market linkages, and once the implications for future investments and capabilities have been assessed, it is essential to examine the fit of the implied strategic path with the firm's existing strategy. Further, the potential for competitive advantage needs to be examined. There are two questions (or groups of questions) to be asked. First, is the investment (and its supporting capabilities) consistent with the existing competitive strategy of the business? Does it introduce requirements that conflict with established modes of operation? In the case of the CHTP for example, a move to open new market niches may create conflict within a marketing organization whose orientation is order-taking and whose incentives push the salesforce toward booking high-tonnage orders for established products. Finding conflict of this sort does not necessarily imply that the new investment should be canceled; it may be time to change the strategy. But asking the question focuses attention on the implication of the decision and on what must be managed well for the innovation to succeed. It may be, for example, that the new process should be set up with a separate management, organization, and competitive mission.

The second question deals with competition. Simply put, does the investment give the firm a defensible advantage over its competitors? Alternatively, if the investment is not made, will the firm be at a competitive disadvantage? The first issue is how hard and how costly it will be for competitors to acquire the same capabilities. A second related issue is whether competitors have or can develop alternative capabilities that will generate a comparable set of product characteristics. Answering these questions requires a careful understanding of the innovation and its implications, both for the firm and its competitors. The firm therefore needs to know what specific advantage the new technology might offer, what specific capabilities are involved, and the nature of its appeal to customers.

But there is more. The firm needs to have as well a transilience analysis of the innovation from the standpoint of its competitors. It needs to know how the new technology might affect competitors were competitors to adopt it and whether any new skills and resources could be acquired at low cost. If the investment provides capabilities that industry participants can acquire easily, any advantage to the innovating firm will be temporary. A defensible advantage is one that competitors will find difficult to better, either because acquisition of the requisite capabilities is inherently costly or because the innovating firm can take action to raise the costs of acquisition. An innovation that does not provide a defensible advantage might appear to yield acceptable financial returns, but it will not alter the firm's competitive position.

Considerations of competitive leverage and competitor action are essential aspects of all four modes of innovation, but the specific issues and likely analytical problems are quite different. In an investment like CHPT, for example, where much of the technology is available for purchase from outside vendors, any advantage will most likely come from the effectiveness with which the technology is implemented and refined for specific applications. The development of supporting capabilities in engineering and marketing, including building customer relationships, may be the important sources of advantage. It is true that the financial requirements may pose an obstacle for following firms, but the development of organizational, technical, and managerial skills in utilizing the technology is likely to provide a more formidable barrier.

In other circumstances—for example, in the case of the continuous micromill—there might be a technological gap between the innovator and subsequent competitors. Here the technology itself may provide the edge that competitors would find difficult to overcome. Further, in this case and in the case of niche creation, there may be advantages to being first, apart from the technological edge. Advantages from reputation, from brand image, from experience, or from preemption of market opportunities might provide a competitive advantage to the innovating firm. Of course, the opposite may be true. If technology is changing rapidly, and if investments are difficult to reverse, followers may have an advantage.

Implications for Practice

The transilience framework has a number of implications for the organization and management of decisionmaking about new technology. First, different methods of analysis, different considerations, and perhaps even a different process are appropriate for different kinds of innovation. Differences in the nature of uncertainty and in the relative importance and complexity of future developments, for example, suggest that a financial framework like simple discounted cash flow (DCF) analysis is not applicable to all kinds of innovation. The simple DCF analysis might be quite useful in assessing the financial implications for a regular innovation where most of the value is in well-defined future cash flows. But a process that focused only on well-defined cash flows might miss much of the value in a niche-creation innovation, not to mention a new technology in the architectural mode. The point is not that quantitative methods are inappropriate in the more complex cases but that simple methods are likely to be misleading.

Where an innovation involves a complex sequence of investments, and where future value is in opportunities to make future decisions, analytical methods that attempt to value those opportunities and assess the probabilities of future opportunities may be quite useful. Where the alternatives are not well defined and the future developments are highly uncertain (as at the birth stages of an entirely new line of technology or business), analysis involving scenarios and technological forecasting may be helpful in clarifying the issues and identifying alternatives. Whatever the particular method of analysis the firm uses, it is important that it be tailored to the problem at hand. And it is essential to understand that no matter what the particular technique, the analysis ultimately rests on the quality of the information about the innovation, about its impact on the capabilities of the firm, and on the characteristics of the product.

And herein lies the power of the transilience framework. It focuses attention on competitive advantage and the capabilities of the firm and its competitors. It reinforces the view that investment in new technology involves the development, nurturing, and replenishing of the firm's productive and creative capabilities. Furthermore, its application requires a thorough understanding of the business and the way that the firm competes and the way that it intends to compete over the longer term. Although rigorous, quantitative analysis can play an important role in the application of the framework to specific innovations, there must be a place in the process for interdependencies, contingencies, and other complexities that may be impossible to quantify. In decisions about the firm's strategic path, there is no substitute for managerial judgment.

But the implications of the framework go further. Not only is managerial judgment (and careful analysis) essential, but the issues that are important demand judgment and insight that cut across traditional functional boundaries. The framework thus has implications for the way decision-making is organized. In decisions about some kinds of regular innovation it may make sense for a staff group to get sales estimates from marketing and cost estimates from manufacturing with little communication between the two groups. In other less stable and more complex situations integration of information and interaction between people with particular expertise is critical.

The framework thus underscores the importance of a general management perspective in decisionmaking about innovation. But it also suggests the value of an articulated competitive mission in each of the functional areas, as well as at the level of the general manager. Without a clear sense of what the competitive task is (that is, what a given part of the firm must do well for the firm to succeed), it is difficult to assess the competitive implications of investment in new technology. Further, if the sense of competitive mission has not been articulated and communicated throughout the organization, the right kinds of innovations may not even come up for discussion.

At the beginning of this chapter I noted the changing competitive environment in manufacturing industries and the pressure on established firms to meet the challenges of international competition and new technology. Meeting those challenges will require vigorous pursuit of competitive advantage through new capabilities in production, technology, and marketing. Aggressive and effective investment in innovation—in new skills, knowledge, and capital—will thus be the key to success in the new industrial competition. And a focus on long-term capabilities, a clear sense of competitive mission, and the seasoning of rigorous quantitative analysis with managerial judgment can be the foundation of effective investment decisions.

4 RESTORING THE COMPETITIVE EDGE IN U.S. MANUFACTURING

Steven C. Wheelwright

If one had surveyed U.S. practitioners and students of management in 1983 as to the competitiveness of U.S. manufacturing firms, there is little doubt that the majority would have responded that the United States was in serious trouble in several major industries. By mid-1984, with the economic recovery gaining steam, a similar survey would have found a much smaller group responding in like fashion, and most respondents would point to the increasing profitability of many companies—even in such industries as automotive products—as a sign of good health.

The speed with which perceptions change is just one of the challenges associated with addressing the topic of a competitive edge in a manufacturing-based enterprise. The purpose of this chapter is not to focus on whether or to what extent U.S. manufacturing has slipped in its worldwide competitiveness but rather to describe what is possible (and required) if U.S. manufacturing firms are to secure a competitive edge in today's environment. However, as background and as a common reference point, it is useful to review briefly some of the available evidence and this author's conclusions from that evidence regarding U.S. manufacturing's current position.

1. *U.S. manufacturing competitiveness has slipped in a broad range of industries.* One evidence of this is the list of industries compiled by Professor Robert Reich (1983) of Harvard University for a Senate subcommittee hearing in late 1980. That list included sixteen industries—automobiles, cameras, stereo components, medical equipment, color television sets, power hand tools, radial tires, electric motors, food processors, microwave ovens, athletic equipment, computer chips, industrial

robots, electron microscopes, machine tools, and optical equipment. In all these industries, according to Reich's research, the manufactured share of the worldwide output produced by United States–based companies fell by more than 50 percent between 1970 and the end of 1979. Clearly, the phenomenon of declining competitiveness is not limited to mature industries. A more thorough and updated analysis of U.S. manufacturing competitiveness and the forces impacting it is provided by Cohen, Teece, Tyson, and Zysman (1984) and in Porter's chapter in this volume.

2. *The decline in U.S. manufacturing competitiveness has not been due primarily to labor costs.* The color television industry presents an important counter to this common theme. As of 1984 there were five Japanese-owned color television plants in the United States (two owned by Sony, one owned by Sanyo, and two owned by Matsushita). Although the number of imports is not insignificant, a majority of U.S. television sales come out of U.S.-based plants, although a sizable portion of those plants are owned by Japanese firms.

3. *The relative competitiveness of U.S. manufacturing may be masked by cyclical swings, but those swings do not change the long-term prospects for such competitiveness.* For example, in the auto industry in the late 1970s and early 1980s, Detroit found itself in a disastrous situation—low volume, high breakevens, and high variable costs. Through dramatic efforts, it was able to reduce their breakeven levels (but without much impact on variable costs), and with the economic upturn beginning in early 1983 and the mix shift back toward large cars, it was able to reap tremendous profits, profits that were higher than those achieved in prior boom years. However, the basic facts with regard to cost per vehicle remained largely unchanged. For a small car, such as the Ford Escort, the U.S.-delivered (to the dealer) cost disadvantage when compared with Japanese competitors was still $2,000 per car as of early 1983. Ford substantiated that through detailed cost analysis in their factories as well as in some of their Japanese joint-venture partners' and through a comparison of market prices where, in early 1983 the market price difference (lower in Japan) for the same car selling in Japan and the United States was almost $1,800. Economic upturns may make the company's bottom line look different, but they tend to have very limited impact on the long-term competitiveness of a given firm or country (Abernathy, Clark, and Kantrow 1983). (The Lawrence chapter in this volume provides additional evidence regarding this point.)

The basic thesis of this chapter is that U.S. manufacturing is in competitive "hot water" in a broad range of industries, labor rates are not the primary cause of that, and a strengthening economy may temporarily mask the seriousness of the problem but does not solve it. The single most important explanation for the worldwide decline in U.S. manufacturing competitiveness is management's view of the manufacturing function, its role, and how that ought to be carried out. Thus restoring that competitive edge requires a basic change in philosophy, perspective, and approach. Such change might be viewed as a prerequisite to the specific strategy and organization changes recommended by other authors in this volume.

This last point cannot be overstated. Widely held U.S. views of manufacturing are, quite simply, incorrect. At best, these hinder progress in manufacturing-based firms; at worst, they lead to continuing decline in worldwide competitiveness.

The balance of this chapter is divided into three major sections. The first describes the primary characteristics of the traditional U.S. view of the manufacturing function, outlines some of the reasons for its emergence, and illustrates several of the management actions that are a consequence of that behavior. The second describes a very different view of the manufacturing function—one that this author believes is a prerequisite to realizing manufacturing's competitive potential. The characteristics, underlying rationale, and resulting management behaviors associated with this contrasting view also are discussed. The third section uses three different types of evidence to support the position that a fundamental change in management's perspective holds substantial promise as a basis for restoring that competitive edge.

The Traditional View of Manufacturing: Static Optimization

The traditional U.S. perspective can best be described as *static optimization* of the manufacturing function. The essential characteristics of this philosophy of manufacturing, clearly evident in the area of *workforce management*, date from the turn of the century when Frederick Taylor (1947) and several of his colleagues initiated what was called *scientific management*. A basic tenet was that management should take more responsibility for how the worker performed the job, that hourly workers should be told what to do, and that their efforts should be controlled closely (often with incentive pay) to ensure that they executed the orders given them. A good description of this view of workforce management is command and control. (This description is very similar to that of military science of that same period, where to *command* and *control* were added the terms *communication* and *intelligence*. Communication referred to the transmittal of data—from manager to worker—and intelligence referred to collecting the data needed to monitor the worker's effort.)

Taylor's description of Schmidt, an hourly worker engaged in loading a rail car with pig iron, is a classic illustration of command and control.

Through systematic data gathering and analysis of Schmidt's activities, Taylor determined the "best" way for him to perform his job. The set of steps comprised by the "best" way was then conveyed to other workers and through close supervisory control, incentive pay, and evaluation, average productivity for the workforce was raised substantially. Although the adoption of such methods in disjointed, almost autonomous departments did improve productivity significantly, the carryover effect on management's views of manufacturing was substantial and not altogether desirable. In terms of the transilience map described by Clark in the preceding chapter regular (retrenching) innovation at a slow pace and totally under management's guidance became the norm.

Basically, management came to view the workforce as a source of energy where eight hours' work for eight hours' pay was the goal. In addition, since white-collar workers and staff were "smarter" (better educated and more highly paid) than hourly workers, it followed that decisions should be made at higher levels in the organization and less discretion left to the shop floor. This increased need for coordination and systems foreshadowed most of today's manufacturing information systems, which in the United States are driven largely by coordination requirements. (For example, material requirements planning systems are aimed at conveying to the shop floor the exact set of actions to be taken, when, and for what products and materials.)

What resulted is management's view that it should be able to plan and decide the best course of action in advance. Implementing this plan is simply a matter of getting the workforce and the first-line supervisors to execute it. This view of workforce management adheres to the oft-quoted phrase, "If it ain't broken, don't fix it." That is, the way to keep things running smoothly in the factory is to stabilize what goes on there, rather than to continually change (and improve) what goes on. Because management is in control of decisionmaking and information dissemination, it is in a position to directly affect such stabilization.

Unfortunately some major negatives emerge from this approach to workforce management. These include a decline in workers' motivation and interest in their work, a decline in suggestions for improvement coming up from the bottom of the organization, the disassociation of workers from overall firm objectives and goals (since it is not their responsibility), and the gradual split into an us-versus-them (workers-versus-management) mentality. Much of today's U.S. labor relations environment is a reflection of this static optimization of the workforce and its role.

Perhaps as a means to accomplish the desired stability in the workplace or simply as a consequence of narrowing the scope of the worker's responsibility, management's perspective on *technology* took on the characteristics of static optimization. This has become particularly apparent in such fundamental aspects of technology as the relationship between product development and manufacturing process development, the role of process improvement (internal to the organization versus external), and the benefits anticipated from more automated production processes.

The static optimization view of manufacturing leads most organizations to view product development as the creative task and manufacturing start-up as the execution task. Thus, new products are developed in sequential fashion with R&D taking full responsibility initially and then tossing that responsibility to manufacturing at a later stage. This connection is often described as the "over-the-wall" transition. R&D does whatever it wants and then throws the design over the wall to manufacturing. With little or no interface between the two functions, manufacturing must figure out on its own what to do with the product designs it receives. This specialization of autonomous functional groups is simply another form of the coordination problems identified by Porter in Chapter 2 as a major hindrance to worldwide competitiveness.

The development of new production technologies is considered of secondary importance. Thus, process evolution in many firms and industries is viewed as an afterthought, lacking the competitive significance of product developments. (It is interesting to note that in petrochemicals and a handful of other process-intensive industries, this view has never dominated.) Processes are developed not in anticipation of new product opportunities but only as required by the pricing realities of the marketplace (costs must be lowered) or the product characteristics as defined by R&D (the existing process cannot be used to make the new product). Because little advance thought has been given to process evolution, manufacturing naturally turns to the apparent experts on such process developments—the equipment suppliers. Thus, the expertise for process technology resides largely outside the organization, and the majority of that expertise must be captured by buying the equipment produced by those suppliers. (The point here is not that equipment is purchased from a third party but that the purchaser sees little need or value in developing technical expertise in the manufacturing processes represented by that equipment.)

Furthermore, this view of process technology asserts that automation is justified primarily as a means of reducing costs. That is, volumes must grow to the point where automation can be justified on a cost improvement basis before it will be considered. Most firms that hold this view make certain that any automation they adopt does not alter the product or its performance in the marketplace. Simply stated, the automated process has to do exactly what the older, less automated process has been doing but at a lower cost. This is the traditional view of substituting capital for labor.

One result of this view of technology and its role in the manufacturing firm is that the anticipated product development and new product performance plans are seldom accomplished. Figure 4–1 summarizes for one firm its expected (before-the-fact) resource allocation and production cost curve and contrasts them with its actual experience on the basis of an examination of its recently introduced products. The results are striking. Product development's anticipated withdrawal from a new product project never takes place because the design as handed to manufacturing is not fully producible. Subsequently, a series of major redesign efforts are undertaken. Although each of these redesigns improves the product, each redesign results in higher costs. Two or three additional redesigns often take place before the product is abandoned and the next generation product is introduced.

In firms with a static optimization view of technology, manufacturing processes lag significantly behind what is possible (that is, their process technologies tend to be old and outdated), and they have much difficulty

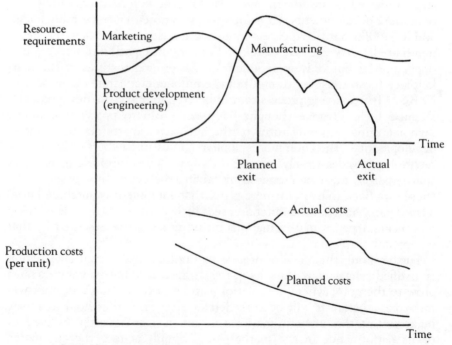

Source: In-company study.

Figure 4–1. Results of Traditional Product Development/Manufacturing Interface

bringing new equipment onstream, complaining continually that the supplier did not tell them enough about what was required to operate it. They find that they have bought equipment from a supplier with no concern for the organizational capabilities required to take advantage of it. Furthermore, because automation is not allowed to affect the product as it is received by the customer, only the most obvious manufacturing technology projects are pursued. Usually, those become ad hoc forays into more advanced manufacturing technologies, rather than a systematic development, evolution, and building of the organization's skills at the most desirable technologies.

This view of manufacturing is largely a *hierarchical or vertical* view that continually subdivides and specializes the function. As a result, it is difficult for the organization to handle tasks that require integration skills, especially if integration is required across functions at multiple levels of the organization. When such coordination is needed, information must flow up to a fairly high level in manufacturing before it can be transmitted either to R&D or marketing. Consequently, many small but cumulatively significant improvement opportunities that might be identified in operations simply are not worth the effort to pursue. Additionally, when new tasks of an opportunistic nature that require an integrated response arise, they are likely to be ignored or given secondary emphasis by the organization because they require capabilities (integration and coordination) that are not readily available. As confirmed by Porter's analysis in Chapter 2, the result is especially apparent in U.S. manufacturing's configuration and coordination difficulties in an increasingly global environment.

Overall, the static optimization view keeps manufacturing primarily in a responsive mode with little opportunity to carry out long-term systematic developments. Thus, although significant effort may still be expended to deliver today's products as well as tomorrow's, the performance never quite measures up to the expectations and manufacturing never becomes a source of competitive advantage recognizable to customers. That may still result in very satisfactory performance, especially when that firm's competitors have a similar perspective on the manufacturing function. However, if such competitors replace this traditional view with a more progressive view that is better suited to today's environment, those adhering to the traditional one will eventually find themselves at a competitive disadvantage.

The Progressive View of Manufacturing: Dynamic Evolution

A much more progressive view is one that can best be described as *dynamic evolution*. In many respects this philosophy is the antithesis of the static optimization view. The dynamic evolution view holds that the problem of production will never be solved. Rather, it is necessary, possible, and desirable to continually improve the function and its contribution to the overall business. This improvement comes not simply from better execution but from ongoing, fundamental change in what is done and how that is carried out. In terms of Clark's transilience map in Chapter 3 a firm whose managers held this dynamic view would find its innovations covering a much broader range of both dimensions of the map, with much higher frequency of occurrence.

To contrast this view with the static optimization view, it is useful to look first at the area of workforce management. Unlike the view of Frederick Taylor, which was one of command and control of *effort*, this view entails investing in the problemsolving skills of the workforce and then focusing their attention on those problems that they are in the best position to solve. This might be described appropriately as knowledge management.

To operationalize this view in an organization, workers' problemsolving skills must be developed and the needed information provided so that they can determine what the problems are and how they ought to be solved. Thus, the concern is not for getting eight hours' work for eight hours' pay but getting the workers' best efforts (both mental and physical) applied to the areas about which they know the most and have the most influence—generally their immediate working area and their production tasks. As described by a colleague, the concern is shifted from *reluctant conformance* under static optimization to *motivated performance* under dynamic evolution. This, of course, requires a very different kind of information system, one that is real-time– and problem-identifying/problem-solving–oriented.

Another characteristic of this view is that since workers have major inputs as to how their work can best be carried out, it is no longer appropriate for supervisors to exercise first-order control—telling them what to do and then measuring to see if they have done it. Rather, second- and third-order control are required, the former consisting of systems and procedures and the latter consisting of shared goals and objectives. The production environment is considered much more worker-dependent than that of the static optimization view and one that involves much more delegation of responsibility, all the way down to the shop floor.

Incentives also must be realigned to reward workers for improvements, to ensure that new learning is captured by the organization, and to make continual improvement, rather than fits and starts, the result. A widely publicized mechanism for carrying out this philosophy of workforce management is quality circles. As practiced by Japanese firms, quality circles are simply a tool that incorporates this perspective on the workforce role into the manufacturing function. Quality circles are not an organizational behavior ploy, nor are they strictly a quality enhancement ploy or a way to keep the workers happy, as they might be if used in a static optimization manufacturing environment. Rather, quality circles are the vehicle for focusing, developing, and applying worker problemsolving skills.

This dynamic evolutionary view of the workforce and its contribution has important implications for middle management. Management's task is no longer the secure one of having all the answers and telling those lower in the organization exactly what should be done. It is much more one of facilitating a team approach, where problemsolving responsibility is shared with workers. For many middle managers accustomed to the static optimization view of the world, this can be a very threatening change in their environment. Similarly, in a unionized plant, such a change has major implications for the union and its role. One thing that can be extremely helpful when a union moves to support the dynamic, evolutionary view of the workforce is for the union to play a major role in developing the technical competence and problemsolving skills of the workforce.

On reflection, most union leaders as well as managers agree that the best way to raise wages is to raise the skills and value added of the workforce. In turn, the best way to increase such skills is through ongoing training and development. It follows naturally in the dynamic, evolutionary view of manufacturing that ongoing training of lifetime employees should be the norm, not the exception. Increasing technical competence then serves as a major basis for worker problemsolving and as a way for management to continually upgrade the contribution of individual workers and work groups.

The characteristics of the dynamic, evolutionary view also can be seen clearly in the area of technology. New product development, which was a sequential procedure under static optimization, becomes truly a team sport. It involves a group of peers who work as equals across functions throughout the duration of the product development/manufacturing start-up activities.

Complementing this view of product development and product technology is the view that process technology is not made up of production equipment alone; it comprises the systems, the people, the equipment, and all aspects of an organization's productive capabilities. Consequently, process technology is viewed as proprietary to the organization even though some elements are acquired from outside suppliers and other organizations. Thus process technology becomes something to be planned for systematically, in parallel with product development activity. The goal of such process development is to provide the capabilities that the product technology will need to support and enhance the latter's translation into distinctive products and customer services.

With product and process technology elements more fully integrated and coordinated, two other aspects of technology management come into sharper focus. One is the fact that over time, and in the majority of cases, significant improvements in technology result from a long series of incremental, usually small, steps and not from a few big breakthroughs. This reinforces the importance of in-house technical capabilities as the key to continued improvement. The other is that shorter development cycles can be achieved, providing three specific types of benefits. The first is better timing of new product and new process introduction. Through product and process coordination such timing can be tailored to the market and competitive realities, and better timing leads to shorter development cycles and thereby to more flexibility either to introduce sooner or to start later and make it possible to incorporate more recent technical developments.

A second benefit of shorter development cycles is improved quality, not only of the product produced (because the designs tend to be more manufacturable and tend to be better suited to the processes available), but also of the basic designs (they are more functional and provide better features). This design quality comes from better focus, better learning, and more responsiveness to the information generated during the product development steps. Basically, with shorter cycle times, those involved in technology development activities can do a better job of capturing the learning that is going on, and there is less forgetting than there is with longer cycle times.

The third benefit associated with shorter development cycles is reduced cost. Reductions in per-unit production costs result from faster more trouble-free manufacturing start-ups, more producible designs, and more flexibility where needed. Costs of the design process also are lowered because fewer steps need to be repeated, overhead on the project is lower when its duration is shortened, and people's time and effort tend to be focused more effectively.

A major plus for the dynamic, evolutionary approach is that, with the emphasis on manufacturing process evolution, automation is viewed primarily as a source of product enhancement rather than simply cost reduction. That is, advanced manufacturing processes are considered a way to produce better products, products that could not be manufactured by others who lack such automation and ongoing process improvements.

The dynamic evolutionary view of manufacturing has other important characteristics with far-reaching competitive implications. One is that as an organization develops this perspective, it significantly enhances its ability to handle what might be referred to as *horizontal* tasks. These tasks cut across functions in the organization at multiple levels. Unlike vertical tasks that can be divided up and handled by increasing levels of specialization, horizontal tasks are considerably more complex and not easily subdivided. A number of tasks facing manufacturing firms in today's environment tend to be of such a horizontal nature. New product development/manufacturing start-up is clearly such a task. Also, the topics of quality and productivity are both of this nature in that they involve multiple functions working at multiple levels to make significant improvements. Similarly, the coordination and configuration tasks in a global environment described by Porter are of this horizontal nature.

Finally, organizations that adopt the dynamic evolution view tend to keep all of their functional areas—marketing, control, manufacturing, and R&D—in much better balance, each making significant contributions to the firm's competitive advantage. Instead of one function (generally either marketing or R&D) calling the primary shots and taking a dominant role in developing the competitive advantage for the entire business, all functions work as a team to develop that competitive advantage. In such firms strong arguments can be made that each of the functions makes a significant difference to the firm's success in the marketplace. The result is an organization that can use its resources more efficiently and that can better combine and integrate their contribution than would be the case where one or more functions is treated as having a "static optimization" role.

Contrasting the Impact of Static Optimization and Dynamic Evolution

There are significant advantages to pursuing the dynamic evolution view in place of the more traditional static optimization view of manufacturing. Although additional direct evidence contrasting these two philosophies and their impact on performance clearly is needed, limited data from three related aspects do lend support to this conclusion. These three areas are those of historical analogy, the experience of a handful of companies that have shifted from one view to the other, and U.S. experiences with the adoption of new manufacturing processes.

From a historical perspective, recent works by Alfred Chandler (1985), Nathan Rosenberg (1985), and Wickham Skinner (1985) indicate that the British and U.S. experiences with industrial development during the late 1800s can be characterized as the British following the static optimization philosophy and the Americans following the dynamic evolution philosophy.

Because of their history and the maturation of their manufacturing firms, the reaction of the British to many of the developments and innovations of the late 1800s was to use them in firms that were narrowly focused, depended on external organizations for technological applications, and had a history of workforce-management confrontation. U.S. firms, on the other hand, tended to integrate technology much better within the individual firm, to address both the worker and management aspects of new technical requirements, and to develop systematically both processes and products that would meet new emerging needs. Chandler (1985: 255) summarizes some of these differences by contrasting the British and U.S. approaches to technology.

> They [the British] failed, not in research, but in development. They failed because they did not create the critical organizational linkages. They failed to build the organization, hire the personnel, and set up the facilities so central to the development of new processes and products. They failed to carry out the developmental processes that are so much more costly in man-power and money than basic research. In particular, they failed to forge the linkages between research, design, production, and marketing so critical to rapid and effective development of new products.

Rosenberg (1985) adds to this description of the U.S. approach the fact that, within companies, within industries, and at a national level, the Americans tended systematically to measure and seek to understand the processes that were emerging. Such measurement became the basis of application for those processes and their continued improvement and refinement. Skinner (1985) adds that because of the pervasiveness of these evolving technologies, it was critical that managers possess the technical competence to interact with those processes and become de-signers of their productive capabilities, rather than simply caretakers of existing manufacturing assets. Skinner describes such people as Andrew Carnegie (in the steel industry) as veritable architects and creators of new opportunities using these evolving process and product capabilities, whereas the British tended to view managers as the owners' agents and thus caretakers of what already existed. Certainly the several discussions in this volume on innovation and entrepreneurship are consistent with this aspect of a dynamic view of manufacturing.

More recent evidence of the differences in impact of these two philosophies can be seen by examining two organizations—General Electric Dishwashers and Mitsubishi Automotive Australia—that have sought to move from the static optimization view of manufacturing to a dynamic evolution view. The first example is the dishwasher unit of General Electric's Major Appliance Business Group. In the late 1970s this business unit, located in Louisville, Kentucky, at GE's Appliance Park, had a twenty-year-old cost-reduced product design, production processes that were ten to twenty years old, a strong union environment of the traditional U.S. adversarial type, and a workforce with average seniority of over fifteen years. However, because of GE's many strengths, this division also had the number-one position in the U.S. market for dishwashers, with approximately a third of the market.

It was at this point that the division proposed to senior management that it be given $18 million of capital to make the next iteration of improvements in the product. Management's response was that if such an investment did nothing to significantly enhance GE's position, perhaps the money would be better spent elsewhere. Division management responded with a serious look at their environment to determine what would be required to significantly enhance General Electric's position. The result was a revised proposal for senior management consideration that included a major overhaul of how the workforce was used in the division, a major new product design, and a new approach to both product and process technology that would upgrade them in an interactive and evolutionary fashion.

This proposal, which had an eventual price tag of $38 million, represented a shift from a static optimization view of manufacturing to a dynamic, evolutionary one. Included in the details of the proposal were changes in the work environment, the skills and participation of the workforce, and management's communication and commitment to them. Also included were a complete redesign of the product with a change from a steel tub and door to a plastic tub and door and an order-of-magnitude improvement in the quality levels (both in the plant and from suppliers). Finally, the production processes were to be significantly altered to incorporate automation that would complement the new product design and deliver the desired quality levels.

This much more aggressive approach was approved by senior management and a three-year effort was launched to change the division's view of its manufacturing function. Within the first nine months of product introduction in 1983, this new dishwasher was a solid success on several performance dimensions, as summarized in Table 4–1. Continued improvement was budgeted (and achieved) for 1984. Additionally, the organization made substantial progress in operationalizing the new view of manufacturing.

The second example comes from the automotive industry in Australia. In the late 1970s, when the Chrysler Corporation found itself in increasing difficulty, it chose to sell many of its "losing" operations. One of those was its auto assembly operation in Adelaide, Australia. That plant had been run under the static optimization philosophy for decades and recently had suffered major losses. Mitsubishi Corp. of Japan acquired the plant from Chrysler and made several significant changes in manufacturing's role and management's view of it, moving it toward a dynamic evolution view. Some of those changes included developing new information systems

Table 4–1.   GE Dishwasher—Results

	1980–81 (actual)	1983 (actual)	1984 (actual)
Service call rate (index)	100	70	55
Unit cost reduction (index)	100	90	88
Number of times handled (tub/door)	27/27	1/3	1/3
Inventory turns	13	17	28
Reject rates (mech/elec test) (percent)	10	3	2.5
Productivity (labor/unit index)	100	133	142

Other: 70 percent fewer part numbers, 20 pounds lighter, worker attitudes (positive 2X, negative 0.5X)

Source: Published GE data (see Hayes and Wheelwright 1984).

(that were problemsolving oriented), eliminating a number of layers in the organization so that managers and workers could interact more closely and more effectively, adopting new management values with regard to worker and manager roles, and implementing a just-in-time manufacturing system that put the emphasis on continued incremental improvement on the factory floor. The results of this approach, contrasted with the last year in which Chrysler owned 100 percent of the plant, are shown in Table 4–2. By 1981 not only had the plant become the most profitable in the Australian auto industry, but its market share had improved significantly (from 9.4 percent to 13.3 percent) and the market share of its primary targeted segment (the midsized automobile) improved from 4 to 34.4 percent!

A third type of evidence of the impact of these two philosophies comes from recent experiences of U.S. manufacturing firms in dealing with "new" manufacturing technologies. Perhaps one of the most highly touted

Table 4–2.   Mitsubishi Automotive—Australia

Results Achieved	1977 (Chrysler-owned)	1981 (Mitsubishi-owned)
Profitability	($27.8M)	$17.7M
Productivity (index)	100	215
Assembly hours per car	59	24
Market share (overall) (percent)	9.4	13.3
Market share (midsized) (percent)	4.0	34.4

Source: McKinsey & Co. study, Sydney, Australia (1982).

technologies on the manufacturing front today is CAD/CAM (computer-aided design/computer-aided manufacturing). Although everybody is talking about it and some firms are adopting it, the results from the use of CAD/CAM are clearly mixed and slow in coming. In one of the most thorough studies of the early 1980s as to the effectiveness of CAD/CAM and what influenced that effectiveness, McKinsey and Company (1984) identified every U.S. user of this new technology and analyzed their approach to the technology and the results achieved. One of the most significant conclusions of their study is summarized in Figure 4–2. They found that adopters of CAD/CAM could be placed into one of three categories:

1. Those who used it simply as a productivity tool for existing individual workers (designers and engineers);
2. Those who used it within a single department (most often the product development group); and
3. Those who used it across multiple functions and at multiple levels.

The overall effectiveness and impact of these three groups of adopters differed significantly, as did the time it took to begin to realize results. As shown in Figure 4–2, those using the technology as a simple substitute for existing processes found real productivity benefits but largely of a cost-reduction nature. Basically, a very powerful tool was being used primarily as an electronic pencil. In the second category were adopters

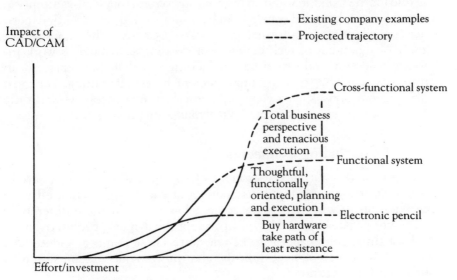

Source: McKinsey & Co. (1984).

Figure 4–2. Approaches and Results of CAD/CAM Application.

who were making major changes in a single department and the way in which that department approached its tasks, but not in the way that department interacted with other parts of the organization. These firms not only obtained cost-reduction benefits (as expected when substituting a newer and better technology for an older one) but also enhanced product features because the new technology provided designers with capabilities not previously available to them. In the third category were those who took a total business perspective on the new technology and who pursued its capabilities across functions and throughout several parts of the organization. For these firms, the technology became a competitive opportunity significantly enhancing the quality of their designs and shortening the product development cycle time. McKinsey found that horizontal integration across tasks at multiple levels was the key difference that distinguished this third group from the first two. Thus, although the first two gained significant benefits, they realized only a fraction of the full competent potential.

A second technology that has appeared on the manufacturing scene in recent years is that of just-in-time production. The basic notion, as originally developed by Toyota, is "elimination of waste." (Such waste covers all activities that do not add value directly for the customer—extra operations, overproduction, defective products, and earlier than necessary operations.) Operationally, substantial emphasis is placed on reducing production cycle times by eliminating work-in-process inventories and shortening equipment set-up times. When combined with a problemsolving role for the workforce and an information system that focuses attention on problems, substantial benefits can be achieved. However, U.S. experiences in adopting just-in-time methods suggest that they consist not simply of a handful of techniques for addressing such tasks as work-in-process reduction and set-up time reduction, but that they represent an entire philosophy of ongoing improvement in manufacturing. For most U.S. organizations, such a view requires a shift in philosophy from static optimization to long-term, ongoing, dynamic improvement.

Concluding Comments

Having read this author's assessment of two very different perspectives on manufacturing and the consequences of each, one might well raise the question of why U.S. management has not seen these differences and responded quickly and broadly to adopt the dynamic evolution view. At least three points appear to be relevant. Those points elucidate why such a shift is necessary yet difficult, and thus provide a useful summary of this entire discussion.

First, when the static optimization view emerged in the early 1900s, it provided substantial benefits to U.S. manufacturing. It recognized the characteristics of the environment at that time—in terms of education level, the shift from loosely affiliated craft groups to large factories, rapidly expanding demand, and opportunities to make order-of-magnitude improvements by adopting and mastering recently developed manufacturing technologies. Because of its success, there was little pressure to adopt a different approach, and management's focus and resources were gradually directed to more pressing areas.

Second, because pressures for a change in that manufacturing view developed gradually and in various forms—none of which "obviously" required such a fundamental change—management could look for other easier, near-term fixes, with some immediate benefits. Thus changes in worldwide competition, domestic market growth rates, the nature of U.S. labor relations, and so forth did not fully combine to significantly heighten pressures on U.S. manufacturing until the 1970s. If anything, the stability of the 1960s and U.S. predominance on the world economic scene caused U.S. management to become overconfident, perhaps even "soft," and miss many of the early warning signals of the need for change.

Third, the shift from the static optimization to the dynamic evolution view of manufacturing is a pervasive and fundamental one. It is not a matter of simply changing a few *decisions* or even making one major issue or *event* the focal point for change. Rather, it requires a change in the *process* of management, encompassing a broad range of behaviors, practices, and decisions as well as philosophies and values. Fundamental change such as this can be accomplished only over years of sustained effort.

The difficulty of making such a change is both the challenge and the opportunity. Although making that change increases the commitment, effort, and resources required to be successful, once mastered it becomes a significant strength, an ongoing basis for competitive advantage. Because it is so difficult to imitate and fits with the environment manufacturing firms are likely to continue to face for the coming decade, it is the best type of competitive edge.

REFERENCES

Abernathy, William J., Kim B. Clark, and Alan M. Kantrow. 1983. *Industrial Renaissance*. New York: Basic Books.

Chandler, Alfred D., Jr. 1985. "From Industrial Laboratories to Departments of Research and Development." In *The Uneasy Alliance: Managing the Productivity-Technology Dilemma*, edited by Robert H. Hayes and Kim B. Clark. Boston: Harvard Business School Press.

Cohen, Stephen, David J. Teece, Laura Tyson, and John Zysman. 1984. "Competitiveness." Center for Research in Business, University of California at Berkeley.

Hayes, Robert H., and Kim B. Clark, eds. 1985. *The Uneasy Alliance: Managing the Productivity-Technology Dilemma*. Boston: Harvard Business School Press.

Hayes, Robert H., and Steven C. Wheelwright. 1984. *Restoring Our Competitive Edge: Competing through Manufacturing.* New York: John Wiley.

McKinsey & Co. 1984. *Restoring Manufacturing Entrepreneurship.* Cleveland, Ohio: McKinsey & Co.

Reich, Robert B. 1983. *The Next American Frontier.* New York: Times Books.

Rosenberg, Nathan. 1985. "The Commercial Exploitation of Science by American Industry." In *The Uneasy Alliance: Managing the Productivity-Technology Dilemma,* edited by Robert H. Hayes and Kim B. Clark. Boston: Harvard Business School Press.

Skinner, Wickham. 1985. "The Taming of Lions: How Manufacturing Leadership Evolved 1780–1984." In *The Uneasy Alliance: Managing the Productivity-Technology Dilemma,* edited by Robert H. Hayes and Kim B. Clark. Boston: Harvard Business School Press.

Taylor, Frederick W. 1947. *The Principles of Scientific Management.* New York: Harper & Bros.

5 COMPETITION: A RENEWED FOCUS FOR INDUSTRIAL POLICY

Paul R. Lawrence

The competitive weakness of many U.S. industries in the international marketplace has not gone away as a major national problem (see Cohen, Teece, Tyson, and Zysman 1985). The economy's recent strength has tended to obscure the problem and also to mute debate on industrial policy. Perhaps the discussion of industrial policy has also waned because it has fallen into the stalemate of little more than a wearisome exchange of familiar arguments from polarized points of view. It is crucial that the debate be renewed so that better ways can be found to improve the competitive strength of U.S. industry.

The discussion on industrial policy has tended to center on the question of whether government should or should not intervene to influence the fate of particular industries. An answer in the affirmative leads quickly to questions of when and how to intervene. A variety of methods ranging from resurrecting the Reconstruction Finance Corporation to creating a U.S. version of Japan's MITI has been proposed. The usual rejoinder to these suggestions is that such planning is in no way compatible with the U.S. political system with its tradition of relying on the push-and-pull of special interest groups. Such critics go on to point to the lack of a competent civil service to carry out federal management of the economy. These arguments are countered by critics who cite the heavy economic and social costs of doing nothing, the cost, that is, of losing the United States' international market position in important industries. As time passes and the arguments swing back and forth, the government is repeatedly forced into crisis interventions (steel and farming are the most recent examples) without having an overall guiding policy that respects both U.S. political traditions and the strengths of the free market. I believe the key to a better industrial policy lies in an historical understanding of the competitive forces that, over time, have been crucial in making industries and companies in the United States economically successful.

The Impact of Competitive Pressure

In a recent, detailed examination of the history of seven major U.S. industries (Lawrence and Dyer 1983), a colleague, Davis Dyer, and I focused on the way organizations in these industries have responded, as organizations, to different kinds and amounts of competitive pressure. We studied the social psychology of competition, the interplay between conditions in the environment and managerial actions that contributed significantly to economic success or failure for these companies. From our examination we realized the basic, but neglected, truth of what I now call the *competitive principle*—

> an industry needs to experience vigorous competition if it is to be economically strong, *either* too little *or* too much competitive pressure can lead an industry to a predictably weak economic performance characterized by its becoming inefficient and/or non-innovative.

How this danger of too little or too much competitive pressure, as seen in the histories of the steel, auto, and coal industries, hospitals, agriculture, housing, and telecommunication is relevant to finding a practical solution to the United States' economic weaknesses is the focus of this chapter.

A middle ground of vigorous competitive pressure in every industry's marketplace, rather than too little or too much pressure, leads management to stress both efficiency and innovation. Although market forces alone can sometimes produce and sustain a vigorous, healthy competitive environment, they do not, by themselves, inevitably do so. Historical instances of government intervention, while always significant in determining the level of competitive pressure, have sometimes been effective and sometimes not. It has been difficult for government to avoid an ad hoc response to each industrial crisis and difficult to avoid being trapped into continuing subsidies. Government intervention has been most successful in moving selected industries away from the unhealthy extremes of weak or cutthroat competition when it was avowedly temporary and carefully designed to restore vigorous competitive pressure.

The history of the steel industry is a good example of the effects of too low a level of competition. From the time of the founding of U.S. Steel in 1901 by the merger of the three largest existing steel companies until almost seventy-five years later, firms in the industry experienced relatively weak competitive pressure. The typical indicators of weak competition were all present: a slow rate of technological advance; high rates of return on investments; high labor costs. Tacit price maintenance, led by U.S. Steel, enabled these conditions to persist. The federal government, by its inaction over many years, was a silent partner in abetting these weak competitive conditions. The emergence of foreign competition in the last several years finally pointed up not only the competitive weakness of most U.S. steel companies but also the consequent high cost in terms of unemployment and community disruption.

The story of the auto industry from the establishment of the Big Three in the mid-1920s to the early 1970s follows similar lines. For many years the competition among the Big Three had been more apparent than real, so that as foreign competition grew, the vulnerability of U.S. firms in terms of efficiency and innovation was exposed. Again the indicators of weak competition were all present.

It is easy for the public to become so accustomed to the ways of an industry that inefficiencies go unnoticed. It took the competitive pressures generated by deregulation in the airline and trucking industries to make the public realize that they could be provided with these services at significantly lower prices. Weak competition, in these cases generated by governmental price regulation, had again been the principal culprit.

Too much competitive pressure has not appeared as clearly in U.S. economic history, but its effects can be as far-reaching and destructive as those associated with too little competition. Although, for the firms involved, the effects of too much pressure are entirely different from those of too little competition, the impact on the wider economy is surprisingly similar. The classic example is U.S. farming during the 1920s. The signs of excessive competition were obvious: a high rate of farm bankruptcies, wildly fluctuating prices, limited gains in productivity, overcapacity, and a low rate of investment in readily available labor-saving and yield-enhancing technologies. All of this was happening while the general economy was booming. Finally, a renewal was triggered in the 1930s, partly as a result of government stabilization of farm prices. A spectacular improvement in efficiency went hand in hand with rapid investment in technical innovations. Since 1929 the workhours needed to produce one hundred bushels of wheat has dropped from seventy-five to seven, a fact that suggests how very sick the industry had been under conditions of cutthroat competition.

The Difference between Competition for Resources and Competition for Innovative Ideas

Distinguishing between the concepts of competition for scarce resources and competition for innovative ideas helps to clarify the issue of the effect competition has on industry. Every firm experiences both kinds of competition simultaneously, but the degree of pressure can vary tremendously between the two types. Roughly speaking, the competition for resources affects the level of efficiency any given organization achieves. When resources are abundant in terms of capital, operating cash flow, people, and raw materials, working for efficiency in utilizing resources often and understandably has a low priority. In this regard organizations respond much like individuals. Neither the social psychology nor the economics of affluence inspire cost cutting. On the other hand, when resources are extremely scarce, efficiency also tends to be low but for very different reasons.

Most economists have argued that in a declining, highly competitive market, the least efficient producers will be forced out of business and the average efficiency will therefore improve. The challenge to standard economic theory comes, however, when some firms have multiple production units. In such situations any marginally efficient firm with but one production unit and significant sunk costs would rationally attempt to survive by pricing as low as marginal costs. At this market price a competing multi-unit firm could be expected to concentrate production of its orders in its more efficient units and therefore close down its least efficient plant, even though it was not the least efficient plant in the industry. Potential new investors also might rationally be deterred for fear that the profits from healthy units become sunk in the costs of closing inefficient units. Furthermore, the social psychology of severe scarcity can lead to many false economies such as the neglect of routine preventive maintenance and employee training—neither conducive to efficiency. Finally, the general reputation of an industry experiencing cutthroat competition makes new investments for improving efficiency in even the stronger units unattractive in spite of the merits of the specific proposal. For all these reasons, under severe competitive conditions, free markets do not consistently and reliably induce efficiency. This analysis, with its inclusion of the social psychology of competition, points to the importance of economists' reactivating their earlier interest in cutthroat competition and reconsidering conventional theory.

The degree of competitive pressure for innovative ideas, in a similar manner, influences the actual rate of innovative action achieved in any given industry. Innovation is a function of information and ideas. Companies, somewhat like individuals, respond to the stimulation of the complexity and change in their environment. Complexity and change create a marketplace for ideas with different degrees of pressure. In principle, innovation may always seem useful to a firm, but slow rates of environmental change simply do not inspire fresh thinking. On the other hand, extreme rates of change and complexity tend to result in an information overload that is also counterproductive to innovative progress. As with resources, an intermediate, lively amount of complexity and change creates the social psychology conducive to innovation.

Some industries can be in the desired intermediate range for one type of competition and not for the other. For example, from about 1950 to 1975, hospitals faced little pressure for resources but lively pressure for adopting new technologies and new forms of medical care. As medical insurance plans mushroomed, it became possible for hospitals to operate, in effect, on a cost-plus basis. Because whatever costs generated were readily passed on to private or public insurers, there was virtually no competition for resources. At the same time, competition for innovation was keen. Hospitals were pressured to adopt the latest health care methods and technologies in order to attract the doctors who would, in turn, bring in the needed number of patients. Innovation consequently shot ahead, but efficiency was neglected.

The coal industry of the 1950s and 1960s provides the opposite example, the kind of pressure for resources that pushes for efficiency, bordering on the extreme, combined with weak competitive pressure for innovative ideas. As a consequence, research and development was neglected; efforts to discover new mining processes, new uses for coal, and new markets were conspicuous by their absence; the industry was low-tech and holding.

Identifying industries in which competitive pressure for both resources and ideas falls within the desirable range is not difficult; retailing, computers, banking, fast foods, microprocessors, and home appliances are only a small sample of those falling into this category. In such industries free market forces can be relied on to foster efficiency and innovation. Competitive conditions in these industries may not be perfect, but in general they are in the healthy middle range. The low-cost leaders in these fields are also actively innovative in improving their products and methods. The innovative first-movers are also cost conscious. Profit margins are usually high enough to permit a continuing investment in new technologies, products, and services but not so high that they foster complacency.

A review of the historical effects of different combinations of competitive pressure in major industries suggests that laissez-faire government policies do not necessarily generate vigorous and healthy competition. Dyer and I found historical evidence of too many instances of either weak or cutthroat competition to be complacent about the automatic working of the free marketplace. One must ask, What determines the amount of competitive pressure in an industry, and how does government figure in the process?

Determinants of Industrial Competitive Pressure

Technical factors have perhaps the most basic influence on the competitive structure of any industry. For example, it is a technical fact that farming depends on arable land and must therefore be geographically dispersed. To carry out farming effectively farm firms must be relatively small and numerous compared to an industry such as automobiles. The inevitable result of the technical need for geographical dispersion is a built-in tendency toward excessive competition in the farming industry. In the electric power industry an opposite situation exists. There, in order to avoid duplication of equipment and service, the technical need for an integrated network of power generation and distribution leads to a natural monopoly. An industry with technical determinants of this sort tends strongly toward having too little competition.

Technical considerations are not static. Technical improvements in transportation and communication have gradually increased the scope of competition from local and regional to national and global in industry after industry. New technologies are not only opening up completely new industries but also changing the competitive structure of existing industries. To return again to the steel example, it was the discovery of an inexpensive way to produce aluminum that created a significant competitive substitute for steel, and it was the development of the electric furnace that made possible the emergence of minimills as significant new competitors to the traditional large integrated firms.

Technical factors not only affect the size and number of viable competitors in an industry with its associated degrees of rivalry, but less directly they affect the nature of an industry's relations with its customers and its suppliers. Again, compare farming and autos: Farmers tend to be squeezed between the large firms that supply equipment, credit, seed, and fertilizer and the large food processing firms that buy many of the farmer's products. Much of the history of farming revolves around the efforts to counteract this technical reality. In contrast, the auto companies by size alone find it relatively easy to dominate their many smaller suppliers as well as the thousands of auto dealers who link to the millions of final consumers. The point is so obvious it is often overlooked: Technical factors vary greatly between industries and create strong tendencies toward weak or excessive competition.

It is only within the context of these powerful technical forces that the actions of business managers can, on the margin, influence the amount of competitive pressure in their industries. When Andrew Carnegie, J.P. Morgan, and others agreed in 1901 to start U.S. Steel with nearly 70 percent of the U.S. steel market, competitive pressure in the industry was quickly and dramatically reduced. But such events are relatively rare. The accumulative effect of many small actions by management, however, can influence the level of competitive pressure in many industries. Having one or two strong industry pace-setters, like IBM in computers, can have an important impact on competitive pressure.

Although government action is only one factor that influences the degree of competitive pressure in any given industry, it is often the most critical determinant because it is the most subject to deliberate control. Government is in the best position to balance and counteract the other influences in the interest of the overall economy and the public welfare. Reviewing the many means of government influence on competitive conditions serves as a reminder of its power and pervasiveness. Antitrust action heads the list of tools the government has at its disposition in affecting competition, but it is only one of many possibilities. Federal tax policies affect competitive conditions in multiple ways, from direct tax credits for research and development, through depreciation and depletion regulations, to the impact of capital-gains taxes on new ventures and dividend taxes on mergers. Trade policies have historically been a favorite to influence competitive pressure. Price supports and price regulation have frequently been used. Direct loans or loan guarantees, direct subsidies, and research grants are all available. The list of government actions that influence competitive pressure includes methods originally designed with quite different objectives in mind. Occupational safety regulations, for instance, have had a major impact on the competitive advantage of coal as opposed to other sources of energy. Environmental regulations affect competitive conditions in uniquely exposed industries such as forest products and chemicals. The impact of defense contracting policies on competitive conditions in many high-tech fields can foster or inhibit the development of whole new technologies. The NLRB affects competitive conditions in many industries by defining the bargaining unit as well as the line between exempt and nonexempt employees. The National Institutes of Health and the Food and Drug Administration have a major impact on the drug and hospital supply industries. The SEC sets some of the competitive conditions for the financial industry, just as the minimum wage and the immigration laws do for many service industries.

I have reviewed this long list to make two points: first, to make it clear that government is already using a large battery of tools that are having a significant influence on competitive conditions in all industries and second, to suggest how largely uncoordinated these many forms of government intervention are. It is, in fact, not unusual for certain federal agencies to be engaged in moderating competitive pressure in a given industry at the same time other agencies are acting to increase the pressure. Our tax laws, for instance, can encourage borrowing for home construction at the same time our monetary policies lead to high home mortgage rates. The Justice Department entered into a consent decree with AT&T to deregulate telecommunications even as new legislation was being enacted to regulate access charges. There are times when a given price list is simultaneously being challenged by one federal agency as predatory and by another agency as collusive.

If it is acknowledged that competitive pressure can be either too great or too small for the well-being of the economy and that government, willy-nilly, plays a critical role in determining that pressure, could the government make its regulatory actions more rational and, in consequence, more effective? I suggest that it could.

Proposed Agency for Monitoring Competitive Pressure

It follows from the argument so far that serious consideration should be given to the creation of a new government agency to monitor competitive pressure on an industry-by-industry basis. Such an agency would use the concept of the competitive principle as a guide in promoting overall industrial health in the United States. Its mission would be to determine and monitor at suitable time intervals the amount of competitive pressure being experienced by each industry. The initial fact-finding and analysis of competitive conditions would be done by the new agency in all industries that have reached some minimum threshold in sales or employment or that otherwise merit federal attention. Regular reports could be issued by the agency describing the competitive pressure for resources and innovation in each industry under study. These reports could guide and advise all relevant governmental policymaking and enforcement agencies in deciding whether they should be trying in each industry to enhance competition, to moderate it, or to leave it alone. These advisories by themselves would help to rationalize and coordinate federal influence on industry. Beyond its fact-finding role, the agency could recommend a coordinated set of governmental interventions as relevant to selected industries.

In industries that are found to be facing excessively weak competition, the agency would have available a wide range of remedies for consideration. The agency could search out hidden federal subsidies that ought suitably to be eliminated. Some form of deregulation might also be indicated; the current wave of deregulation has already dramatically altered traditional assumptions that natural monopolies require extensive regulation. The simple act of spotlighting weak competitive conditions would most likely encourage the entry of new competitors from related industries. In other words, with all the moderate tools that government has available for enhancing competition, only in exceptional cases would it be necessary to fall back on antitrust action, the ultimate government instrument. When the proposed agency finds that a particular industry is clearly suffering from excessive competition, it could consider the use of a long list of possible remedies that have been employed by the government at one time or another. Trade quotas, fast depreciation allowances, and loan guarantees are only the currently popular methods. Historically the government has relied more on tariffs, price supports, and fair trade rules. The list could easily be extended. Good policy in this regard depends on finding the right combination of actions to achieve the desired "cure," always minimizing the adverse side effects and always being particularly careful not to create a lasting dependence on the remedies employed. The medical metaphor is deliberate, for, as with drugs, government remedies can be overdosed as well as underdosed. And as with physicians prescribing drugs, government policymakers must recognize that efforts to moderate the symptoms of disease should last only until the patient is restored to health. Continuing support indefinitely in cases of cutthroat industrial competition leads inevitably to the opposite problem of too little competition—such as in programs designed to reduce extreme agricultural competition that have been continued for so long that they now cause serious problems in certain agricultural areas such as dairy and tobacco farming.

The proposed agency should be prepared to identify and act in regard to situations when competition for ideas is excessive even when competition for resources is in the healthy range. There have been cases in which the government has shown the capacity to distinguish different forms of competition and use appropriately different tactics. A case in point is recent governmental action in the semiconductor industry, in which competition for the development of ideas has become excessive and wasteful at the same time that the degree of competition for resources is appropriate. A number of U.S. semiconductor firms, feeling extreme competitive pressure, have invested large sums in R&D to solve the same problem in memory-chip design. To avoid this duplication of effort, the government is now willing to legitimate cooperative research and development among these firms without jeopardizing their healthy competition for resources. To build a healthy economy, government needs to encourage managers in an industry to work together in some areas even as they remain scrupulously apart and compete vigorously in others.

Hazards of Government Intervention

To propose the creation of a government agency to oversee intervention at the industry level is to propose the use of a two-edged sword. The monitoring and maintaining of the right amount of competitive pressure can be useful, but it must be done with great care. Success will require constant vigilance and careful attention to the particular needs of each industry. At times obvious competitive indicators can be misleading. For instance, although IBM's dominant domestic market position, more than any other single indicator, seems to have prompted the Justice Department to launch its antitrust investigation of the company, it is now apparent that the government did not take sufficient notice of the build-up of global competition in computer technology. Also, because IBM has always been successful in creating among their employees a strong sense of external competitive pressure, they have consistently been innovative and efficient. Government intervention was, in fact, neither necessary nor helpful.

The greatest hazard for an industrial policy agency is that of creating a continuing dependence on governmental aid. Shipbuilding, for example, in the name of national security, has been heavily dependent on federal subsidies since World War II. Although wartime innovators like Kaiser developed highly original, low-cost methods of shipbuilding, traditional U.S. shipbuilders, already addicted to subsidies, showed no interest in adopting these new techniques; Kaiser consequently taught them to the Japanese. Now, some forty years and billions of subsidy dollars later, the United States has a weak shipbuilding industry that, in spite of recent efforts to wean it from governmental dependence, is far from matching foreign competition. There are, of course, strong reasons for producing our naval vessels in U.S. yards, but our continuing subsidy of merchant vessels has not worked to our advantage.

The Chrysler case, on the other hand, is an example of the government's intervening with financial help in a most successful way. This success was no accident. The so-called bail-out was carefully designed to make the guaranteed government loans contingent on remedial steps to be taken by the company and other key parties: the union, creditors, suppliers, and dealers. All parties agreed to a tough, realistic plan for the recovery of the firm's competitive strength. Such a hard-headed, quid-pro-quo deal should be carefully built into every plan for government assistance. Unfortunately, there are signs that the federal government is already forgetting the lessons to be learned from the Chrysler case: The government provided voluntary import quotas to help the entire U.S. auto industry but failed to require a corresponding work-out plan from the auto firms. The same mistake seems to be happening as regards the steel import quotas.

Were the competitive principle explicitly endorsed by government, any government effort to moderate competition would be viewed as a unique and temporary arrangement, solely designed to restore normal competitive pressures. Special pleading and lobbying will, of course, continue but their impact could be blunted. General adherence to the competitive principle could significantly reduce the number of cases of unhealthy dependence on government aid.

The final major problem area in adopting the competitive principle is that of measurement. Although measuring, on a continuing, industry-by-industry basis the amount of competitive pressure for resources and innovation will not be simple, the task ought not to be any more difficult, for instance, than those regularly tackled by the Federal Reserve Board.

Implications of the Competitive Principle for Corporate Management

Government application of the competitive principle in its industrial policies would provide a boost to business performance that is hard to overestimate. With government providing consistent and understandable ground rules, business would be able to move ahead with longer-term resource commitments that are now held up by governmentally generated uncertainties. In recent years the federal rules have been in a constant flux in terms of taxes, trade policy, monetary and fiscal policy, antitrust, and price regulation. This uncertainty has caused a constant planning and replanning, a hesitancy about making major investments, a constant preoccupation with lobbying and politicking. For business leaders to ignore these uncertainties is suicidal, but responsible managers have deeply resented these distractions from running a truly efficient and innovative enterprise.

Top managers have been forced to spend a staggering amount of time on governmental affairs. The cost in terms of the ever-increasing services of corporate lawyers and regulatory affairs officers has been enormous. Redirecting such time and resources back to the nuts and bolts of the business would, in itself, be a step toward improving the competitive strength of U.S. business. Those business people who have found satisfaction in spending their time in the corridors of Washington will be disappointed, but most managers will respond with a sigh of relief.

To achieve these benefits, however, business leadership must face how the competitive principle affects their behavior. Application of the principle generates some do's and don'ts about how to compete and how to manage their organizations. The resulting rules of conduct need to be understood and accepted.

Certain methods of competition would need to be embraced and others rejected. Management would need to focus on earning profits by being efficient and innovative instead of by trying to lock-in profits. Earning profits means developing distinctive corporate competences. Such special competences can take many forms: evolving particularly effective methods of manufacture and distribution; developing uniquely useful products and services; tailoring product or service to special customer needs. Such competences can be distinctive in the sense that they are visible and valued by the customer and also distinctive in being difficult for competitors to imitate. In other words, some modest frictions in the free marketplace are healthy inducements to longer-term investments in people, ideas, and equipment. On the other hand, searching for ways to lock in profits is not compatible with the competitive principle.

Just as there are many ways to develop distinctive competences, so there are many ways to lock in profits; tacit price maintenance is probably the most important one. We have already cited the consequences of long-term tacit price maintenance in the steel industry. When Morgan selected Judge Gary to head the newly formed U.S. Steel, Gary quickly spread the word that, as long as no one rocked the boat, U.S. Steel's established price schedule would virtually guarantee comfortable profits for everyone. These understandings persisted with amazingly little change for decades—until overseas competition exposed the seriously weak competitive state of the industry. The point is obvious: Mergers and other methods that enable tacit price maintenance violate the competitive principle, and managers who go outside these bounds should expect prompt governmental countermeasures.

The other significant category of behavior violating the competitive principle is created whenever an industry or firm becomes basically dependent on government subsidies for its survival. Businesspeople have been guilty of finding many ways to rationalize continuing to receive government support; they repeatedly have lobbied for continuing government subsidy in the cause of saving jobs, protecting the community, assuring national security, or whatever. Under conditions of truly cutthroat competition, in accordance with the competitive principle, a business can legitimately ask for temporary government help to moderate competitive pressure but must expect to present a realistic and expeditious plan for restoring normal competitive conditions.

Firms facing healthy, vigorous competition are continually challenged to solve the inevitable conflicts between efficiency and innovation. Maximum efficiency calls for uninterrupted production to take advantage of the learning from experience and the economies of scale. The push for innovation calls for continual experimentation, searching for new and appropriate innovative procedures. The two goals are inevitably in conflict, with the cultural style of a creative R&D shop very different from that of a high-volume, low-cost manufacturing unit. The basic tension between efficiency and innovation resurfaces in a variety of different forms: short-term versus long-term planning, control versus entrepreneurship, tradition versus creative change. Although one side of these dichotomies may be emphasized, neither side can be allowed to dominate. The dominance of efficiency results in outdated products and processes; the dominance of innovation becomes too costly.

Finally, firms facing vigorous competition usually have found it necessary to involve all ranks of employees in meeting the competitive challenge. Such firms are finding that they can no longer afford the adversarial relations with labor that have characterized many of our industries in the past. As competitive pressures mount, firms are moving toward shared management in an all-hands approach to problems. To increase organizational effectiveness they are sharing information, sharing in solving problems, sharing rewards and sacrifices, and even sharing power. Manufacturers of automobiles, for example, have learned that they cannot prosper if workers on the assemblylines show up every day with hostile or apathetic attitudes toward the company. High involvement of employees through shared management is likely to become a competitive necessity.

The Advantages of an Explicit and Flexible Use of the Competitive Principle

The competitive principle for guiding federal intervention at the industry level has four main general strengths. First, it is simple and straightforward and fits easily into the U.S. ethos. If government's goal is vigorous competition, it is clearly rational to step up the competitive pressure when it is too little, to moderate competition when it is too much, and to leave it alone when it is, in fact, vigorous. The approach is built on the entrenched U.S. belief in the value of healthy competition and the conviction that market forces normally produce the most efficient allocation of resources through free consumer choice. It also acknowledges the hard fact, however, that some industries, at times, approach conditions that border on monopoly or, at the other end of the spectrum, the chaotic conditions of excessive competition. In either case, the carefully designed and temporary intervention of the government is called for to restore normal competition.

Second, the competitive principle conforms to our traditional political practices and philosophy. We do not expect the government to pick the industries that will survive while selecting others for extinction. Such government choices would violate our historic understanding that the government defines and polices the ground rules for competition but does not determine the winners and losers; that companies have the freedom to fail as well as to succeed. The United States cannot arbitrarily write off an entire industry as a matter of government policy. The United States' market with its size and diversity has the potential to sustain almost every industry at some level, in some form. Industries are dynamic, always changing: Some labor-intensive industries (such as textiles and autos) might renew themselves by becoming more capital-intensive; some low-tech industries (such as trucking and steel) show signs of renewal by becoming high-tech. No government agency can predict these changes accurately, much less initiate them. Instead, government's job should be confined to the much simpler task of identifying whether appropriate competitive forces exist in different industries. This limited form of industrial intervention differs from both the French variety, with its implicit allocation of market shares, and the Japanese, with its implicit allocation of capital. The types of intervention we propose have been practiced on an ad hoc basis in the United States since its founding (see McCraw 1986): What is new is the explicit formulation of both a guiding principle and of the reasons for adopting it.

The proposal is flexible enough to accommodate significant industrial change and adaptation. History shows that the level of competitive pressure in any given industry changes over the years but that such changes have been difficult to predict. However, the administrative agencies of government should be able to respond after the fact to these changes and respond without the delays inherent in new legislation and court proceedings. Economic affairs are too complex and dynamic to expect any one set of government rules to be suitable for all industries and over long periods of time. The proposed guidelines for government intervention recognize that different industries will require different remedies to sustain an underlying consistency of purpose—maintaining vigorous competition.

Finally, the proposal is feasible. The Federal Reserve Board, with its task of using monetary measures to steer the entire economy between the extremes of boom and bust, provides a close analogy. Its performance is hardly perfect. We still have swings. But it is constantly learning to perform better, and its record in moderating cycles is a clear improvement over the past. An industrial policy agency should in time be able to perform as well. Once in place, such an agency would be in a position to implement a national industrial policy that could endure political swings; a policy that could increase the number of U.S. industries capable of competing in the world market. The economic adversities of the past decade have renewed widespread respect among Americans for the value of healthy competition. It is now the time to build on this political fact and adopt the competitive principle as the foundation of our national industrial policy.

REFERENCES

Cohen, S., D. Teece, L. Tyson, and J. Zysman. 1985. *Competitiveness.* Published as Vol. III, *Report of the President's Commission on Industrial Competitiveness.* Washington, D.C.: U.S. Government Printing Office.

Lawrence, Paul R., and Davis Dyer. 1983. *Renewing American Industry.* New York: Free Press.

McCraw, Thomas K. 1986. "Mercantilism and the Market: Antecedents of American Industrial Policy." In *The Politics of Industrial Policy*, edited by C.E. Barfield and W.A. Schambra, pp. 33–62. Washington, D.C.: American Enterprise Institute for Public Policy Research.

Part II

The Theoretical Context of Strategic Management

6 BRINGING THE ENVIRONMENT BACK IN: THE SOCIAL CONTEXT OF BUSINESS STRATEGY

Jeffrey Pfeffer

The academic field of business strategy or strategic management has developed in a fashion consistent with many of the other branches of U.S. social science that are focused on understanding organizations, particularly business organizations. The disciplines of economics, political science, organization theory, psychology, and even sociology, for the most part, have been characterized by two perspectives that are at once both taken for granted and problematic for furthering our analysis of organizations: (1) an analytical approach that Baysinger and Mobley (1983) characterize as methodological individualism, in which the individual is the unit of analysis, or if larger social aggregates are being studied, they are analyzed as if they were individual actors; and (2) a focus on rational action as the explanation or prediction of action taken by these individual social units. As Granovetter (1985) has noted, there has already been extensive critique of the logic of rational action and, moreover, the assumptions of rational models of choice are powerful both in generating predictions of behavior and in serving as parsimonious explanations of empirical observations.

This chapter critiques the approach to analyzing action that is based on methodological individualism. First, it briefly reviews the existing approaches to strategic management, indicating how all such currently popular approaches fail to pay attention to the relational, social nature of organizational life. Then it reviews evidence indicating the importance of interorganizational power and consequently why interorganizational power should be a (if not *the*) focus of strategic management and strategic action. Next, it considers what we know about the development and explanation of variation in interorganizational power. Finally, the chapter concludes by demonstrating how this perspective, rooted more firmly in the realities of the social nature of organizational and interorganizational life, both broadens and alters the conception of strategy and the types of studies of strategic management one might undertake.

The Existing Strategy Literature

Without reviewing the strategy literature in great detail, let me note that the literature is often, and I believe, usefully categorized as follows. In the first instance, there is a distinction between the content aspects of strategy, what the firm should do in terms of what markets to be in and how to approach those markets, and the process aspects of strategy, most often encapsulated in the idea of issues of implementation. Within the content domain, there is often a distinction made between corporate strategy, which focuses on what markets or businesses the firm should be in, and business-level strategy, which "focuses on competition within particular product/market segments" (Astley 1984: 528). Thus, corporate strategy directs attention to what environments to operate in, and business-level strategy focuses on how to compete effectively within those environments—for example, through pursuing strategies based on economies of scale and low-cost production or, alternatively, product differentiation and service.

It is fair to state that both corporate and business-level strategy research has, to this point, been characterized by a fundamentally internal focus with the single organization as the unit of analysis. Although it is clearly true that no strategic research or practice fails to take into account the nature and characteristics of the environment facing the firm, such as the number of competitors, market shares, the dimensions of competition, and so forth, it is also the case that in most research and practice in the domain of strategy, these external constraints are taken as given, as characteristics of the environment to which the firm must adapt in order to be successful. Thus, as Astley (1984: 526) has noted, "organizations are viewed, basically, as solitary units confronted by faceless environments." It is in this sense that we can speak of an internal focus of strategy and strategy research. Astley and Fombrun (1983: 576) have noted that "Strategic action therefore is characterized in terms of a predominantly internal focus, a concern with matching organizational capacities to environmental demands." The focus is internal in that action is directed internally, in lowering production costs, in integrating to absorb sources of supply and lower transaction and production costs, in altering product positioning and product development and introduction strategies. Although each of these actions affects the environment, the actions are taken within the organization's boundaries and are, in this sense, internal.

That the organization is viewed as an isolated unit is, perhaps, more obvious. Strategy research is written from the point of view of the focal organization. Even analyses that presumably focus on population or industry-level conditions—such as population ecology (Hannan and Freeman 1977) in organization theory and competitive industry analysis in economics (Porter 1980)—really do not implicate the structure of relationships among organizations or the embedded, situational character of relationships among units in their analyses. In both instances, environments are characterized in terms of niches, resource pools, and the dynamics of competition but in terms of an undifferentiated mass of faceless competitors who are at once both not proximately tied nor, at times, even presumed to be responsive to the actions of the focal organization. Thus, industry conditions, or niche conditions, become environmental attributes or dimensions to be included in the analysis as factors to which the focal organization is presumed to adapt internally or, in the case of ecology, as dimensions along which the organization will be selected depending on its fit.

The process approaches to strategy, dealing with issues of implementation and strategy formulation (Hrebiniak and Joyce 1984; Mumford and Pettigrew 1975), are even more focused both internally and on the organization as the unit of analysis. The issues confronted include ones such as whether or not strategy—in the sense of rational, conscious, foresightful planning—is possible under the pressures of day-to-day activities and internal political jostling (Bower 1970) if strategies are developed, how can they be communicated in such a way as to be implemented in the myriad decisions made not only at corporate headquarters but in the business units themselves; and how the planning process itself can be best organized and conducted, including the accessing of relevant expertise and the involvement of line managers.

These approaches to understanding both the content and process of strategy have been useful but are clearly incomplete. Their incompleteness is difficult to comprehend because of the pervasiveness of the emphasis on methodological individualism in both social science and social thought. Their incompleteness comes from the fact that in a way analogous to individuals within organizations, organizations themselves exist as part of networks and systems of other organizations. Relations among organizations are embedded, to use Granovetter's 1985 term, and they are embedded in the sense that they have a history (almost invariably neglected in all the formulations present in the literature) and a structure, also always neglected. Just as organizations are structured, systems of organizations are structured, and there exist institutional elements of these structures that need to be attended to in doing either research or practice in the area of strategy. But before getting to this point more explicitly, perhaps we can motivate the argument even more by considering the evidence on the relationship between power and profits.

Interorganizational Power and Profits

In both economics and sociology, there has been continuing interest in the effect of power on profits, although as is noted below, other dependent measures beside profit might be more appropriate. In the case of industrial organization in economics, the studies have examined the effect of market power, first measured by things such as the concentration ratio and then by more refined measures of market structure such as the Herfindahl index, on outcomes such as the price-cost margin, profitability, and the excess of the firm's market to book value (see Weiss 1963; Collins and Preston 1968).

In sociology and organization theory, this work is exemplified by Burt's (1980) analysis of the relationship between structural autonomy and profit. Burt (1980: 895–96, 899) argued that there were two aspects to autonomy:

> one aspect of autonomy concerns the relations among actors jointly occupying a status in a system. . . . The actors . . . will be able to escape the constraints of supply and demand imposed by actors in other positions and, accordingly, will be "autonomous" within their system, to the extent that among persons, or corporate actors, occupying the position there exists an oligopoly . . . or, in the extreme of centralization, a monopoly.

> a second aspect of autonomy concerns the manner in which actors jointly occupying a status are related to actors occupying other statuses in their system. . . . Actors . . . will be able to balance demands from other actors, and accordingly, will be "autonomous" within their system, to the extent that the pattern of relations . . . ensures high competition among those actors who interact with the occupants of position . . . a measure of autonomy via group-affiliation must consider two things: the extent to which actors occupying a status have diversified relations with other statuses, and the extent to which they have relations only with statuses that are too poorly organized to make collective demands.

In other words, power in a network comes from having centralized control or coordination over those similarly situated in the network and from dealing in a more diversified way with the rest of the network and particularly with other groups that are not organized or centralized.

Burt (1980) found that his measure of structural autonomy was related to profit margins at the industrial sector level of aggregation. Of course, as various economists (such as Caves 1970) have noted, firms may choose to take some of the benefits that accrue from positions of power in things other than profit, such as reduced variation in performance or, in other words, more certainty and stability. One useful extension of Burt's work would be to replicate his results using other dependent measures besides the one he used, including variation in profits as well as measures of performance and risk more directly tied to security prices. Indeed, given an appropriate sample in which there was mortality observed over the time period of the study, one would predict that Burt's measure of structural autonomy would be related to survival prospects, with more autonomous organizations being more likely to survive.

Two elaborations on the Burt study are important. First, from the point of view of strategic management, Burt's methodology enables one to, given the availability of the necessary data, find positions in transaction networks in which there will be more autonomy and hence greater likelihood of earning a higher return. Thus, Burt's network formulation can guide the choice of market environments for the firm to be in. Second, and possibly more important, note how Burt's framework, or any other that would explicitly incorporate network thinking, differs from other formulations. For instance, in the Freeman and Hannan (1983) study of restaurants, a given restaurant's likelihood of survival was predicted by its form (specialism or generalism) interacting with the conditions of the environment in terms of grain and variability. No consideration is built into the analysis of the restaurant's patterns of relations with suppliers, customers, or competitors in the area. Yet no study of the retail industry is really complete without such factors being included. In the grocery industry, for example, wholesalers often furnish the capital for retail stores (independents) to modernize, and relationships with wholesalers affect the availability and cost of credit as well as the goods actually sold. In the motel and, I suspect, the restaurant industry, network relations are developed with client organizations such as corporations (in the case of motels) and corporations and organizations (in the case of restaurants) to ensure to the extent possible a steady base of business. One of the most important determinants of the profitability of motels in the Palo Alto area is whether or not the motel has developed a stable client relation with some set of local businesses. Although the motels may look alike, or may have similar characteristics in terms of their form, their network positions and consequently their financial success differ vastly. It is exactly this distinction that demarcates, in my view, approaches to analyzing strategy that are fundamentally individualistic in their orientation to those that are more explicitly relational and structural.

If profits and possibly survival and stability are related to the position of interorganizational power, then it makes sense to argue that at least one important objective of strategic management should be to enhance the firm's interorganizational power. And accepting that premise moves us even further into considering the relational aspects of organizational life.

Determinants of Interorganizational Power

A number of theoretical perspectives address the issue of the determinants of interorganizational power. One, the structural autonomy approach of Burt (1980, 1982, 1983), has already been mentioned. According to Burt, power accrues to those actors that occupy positions with other actors that are centralized or coordinated and deal with a diversity of other sectors that are themselves uncentralized and unable to engage in coordinated action. This perspective suggests that actions taken that (1) diversify the firm's dependencies, (2) direct transactions to less organized or centralized sectors, or (3) facilitate coordination within the firm's own sector, should enhance power and consequently those outcomes that follow from power. Following this logic, what is critical about diversification is not necessarily product diversification but, rather, diversification that causes the firm to depend on less concentrated and more dispersed sources of supply and customers. Providing products that require essentially the same inputs to similar markets would not constitute an increase in the firm's power. It is particularly critical to diversify away from concentrated sectors. This is consistent with the finding reported by Pfeffer (1972) that firms that tended to transact more with the government tended to engage more in mergers for diversification. Accomplishing the third part of the strategy, coordinating activity within the sector, might be achieved either through horizontal mergers to absorb competitors or through other forms of coordination—such as trade or industry associations, interlocking directorates, or joint ventures—or having coordination imposed by the government as through regulation.

Another perspective on interorganizational power comes from network analysis. Actors that are more centrally located and more interconnected should have more power. Note that this approach leads to a somewhat different set of predictions, as structural centrality is certainly not the same thing as structural autonomy. According to this approach, firms or sectors that are more central in exchange relations should have more power. Mackenzie and Frazier (1966), some years ago, measured structural centrality in the network of transactions associated with a wood products market. Following the argument that centrality is related to power and power to profits and other such outcomes, one could use his methodology to study other markets as well as the consequences of centrality in those markets.

Both structural autonomy and network approaches focus on transactions as they occur. But transactions and the discretion to use the power that emerges from networks of exchange relationships are both under the partial control of other authority such as the government. This institutional structure, as well as power emanating from patterns of resource exchanges, is somewhat better captured by resource dependence theory (Pfeffer and Salancik 1978: 75). This approach specified ten conditions that would affect whether or not a given organization would comply with external demands and, by extension, its degree of autonomy or power:

1. The focal organization is aware of the demands.
2. The focal organization obtains some resources from the social actor making the demands.
3. The resource is a critical or important part of the focal organization's operation.
4. The social actor controls the allocation, access, or use of the resource; alternative sources for the resource are not available to the focal organization.
5. The focal organization does not control the allocation, access, or use of other resources critical to the social actor's operation and survival.
6. The actions or outputs of the focal organization are visible and can be assessed by the social actor to judge whether the actions comply with its demands.
7. The focal organization's satisfaction of the social actor's requests are not in conflict with the satisfaction of demands from other components of the environment with which it is interdependent.
8. The focal organization does not control the determination, formulation, or expression of the social actor's demands.
9. The focal organization is capable of developing actions or outcomes that will satisfy the external demands.
10. The focal organization desires to survive.

Note that interdependence, in the sense of the exchange of resources, is the focus of only two of the ten conditions. Other conditions focus on the ability of external agents to enforce demands (that is, for price, quality, or quantity), the nature of external actors' dependencies on organizations operating in the sector, and the visibility of the response of the organization or set of organizations. As noted in their analysis, power that emerges from resource exchanges is often proscribed on the one hand, and other forms of power are derived from governmental and other institutional actions and arrangements.

Some Implications for the Research and Practice of Strategy

The implications of the preceding argument for the study of strategic management can be articulated at several levels of generality. At the most general, it suggests an appreciation of the concept of "collective strategy" (Astley 1984; Astley and Fombrun 1983). Collective strategy involves "collaboration, or joint action by organizations on matters of strategic importance" (Astley 1984: 524). As Astley and Fombrun (1983: 577) have noted, "in a corporate environment characterized by increasing interdependence and ever more intricate networks of linked organizations, individual strategies are overwhelmed by proactive choice at the collective level. . . . there is the increasing emergence of structures of collective action, ranging from informal arrangements and discussions to formal devices such as interlocking directorates, joint ventures, and mergers."

Hirsch (1975), almost a decade ago, taught us that in spite of numerous similarities in the characteristics of the product and the fact that both were essentially distributed through intermediaries, there were vast differences in profitability between firms operating in the pharmaceutical and the record industry, and indeed, recent evidence indicates that the pharmaceutical industry in different countries varies tremendously in its profitability. Hirsch's analysis points to the differences in profitability deriving from conditions of the institutional environment faced by the two industries, including restrictions on product entry, pricing, and promotion that are mandated by legislation and regulation. Hirsch's work suggests that understanding strategy in the two industries should begin by considering what the drug companies have done to both attain and maintain a favored competitive position and how they have done so. Many of these actions are cooperative and collaborative; most involve activities well beyond diversification, integration, and product market positioning, which constitutes the focus of more conventional strategic analysis.

Hirsch's analysis and argument also teaches us that there is probably more variation in profits or other measures of performance between industries than within. Ironically, most of the research and thinking on strategy seeks to achieve competitive advantage or compare performance variation across individual units. But this is not where most of the variation or the strategic action is. As Hirsch argued, the critical factor may be the particular sector or institutional environment itself and how and why it came to be structured the way it is. Thus, Burt's analysis of intersector variation in profits may be couched at precisely the right level of analysis, and the idea that strategy be analyzed at the sector or industry level of analysis is similarly useful.

As another example of collaborative strategy, consider the automobile industry. Two conditions stand out in this industry. First, it is likely that the recent robust profits are at least partly the result of import restrictions placed on the Japanese. Such restrictions were obtained through a coalition of automobile companies and autoworker organizations. Second, without exception all of the U.S. manufacturers are involved in joint ventures with either Japanese or European automobile companies. Furthermore, this practice of joint venturing now extends to some segments of the computer industry, to aerospace, to farm equipment, and Hall (1984) has argued that joint responses to industry changes is the coming modal strategic response. If such is the case, issues of coalition formation and negotiating the terms and participation in these interorganizational organizations will become increasingly important activities.

On a somewhat less general level, the focus proposed here implies attending to a range of strategic actions and responses that is considerably broader than that traditionally included in the rubric of either business or corporate strategy. Such activities include not only mergers to reduce both competitive and symbiotic interdependence (vertical integration), but also joint ventures, board of director interlocks, and perhaps most important, political activity of all shapes and varieties. The merger of subject matter occasionally occurring in business schools between the fields of business and society and strategy is more than serendipitous—it is essential given the critical role of political action in the context of corporate and business strategy.

And finally, on the level of specific research hypotheses and directions, numerous areas for study are suggested. If power accrues to those sectors of the economy that are able to coordinate and organize, we need much more systematic study of what Phillips (1960) termed *interfirm organizations*. Pfeffer and Leblebici (1973) suggested that the movement of executives across firms may facilitate interfirm coordination, but they investigated only some of the determinants, not the consequences, of such interfirm movement. Similarly, Pfeffer and Salancik (1978) argued that interlocks among directors of competing or potentially competing organizations facilitated coordination, but, again, the causes, not the consequences, of such patterns of relationships were partly assessed. Burt (1983) has investigated whether firms that attempted to coopt constraining sectors were more profitable, finding little effect. However, this may be for two reasons. First, as Burt (1983) noted, most of these interorganizational linkages already followed the pattern of constraints. More important, Burt's analysis tended to focus on coopting relations *across* sectors, rather than on organizing and coordinating activities accompanied by interfirm activities *within* sectors. An expansion of his analysis to incorporate the possibility of intrasector coordination as a predictor of profit and other outcomes would seem to be warranted.

We have already noted that network position might be used to predict survival as well as profits at the firm as well as the industry-sector level of analysis. The more structurally autonomous the position, the higher should survival prospects be. But ecology concerns birth as well as death, and it would be interesting to examine whether there is also more entry in structurally autonomous network positions than in positions in transaction networks that are more constrained. If Burt's analysis is correct and profit is higher under less constraint, and higher profit stimulates entry, then birth as well as death should be at least partly predictable from network characteristics.

However, entry into an industry depends on the stability as well as the pattern of network relationships. Consider a firm that is thinking about or actually trying to enter a service supplier segment of the construction industry, such as painting, plumbing, or electrical work. Certainly at times, the apparent monetary rewards look large. Furthermore, there are many general contractors and many suppliers of services, so that the industry would appear to be competitive. With limited capital required to enter many of these lines of work, and with the technology fairly rudimentary and stable, there should be both rapid and successful entry, particularly in times of high demand. I suspect there is not because as Eccles (1981) has noted, in this industry as in many others, apparently independent entities are actually organized and structured as quasi-firms. In this instance, each general contractor does business with one or only a very few subcontractors of each type, and these relationships are stable over time. Thus, this apparently diffuse and amorphous industry is, in fact tightly structured—so tightly structured that entry may be difficult, particularly for firms entering when there is not a corresponding increase in the number of general contractors so that success becomes dependent on actually disrupting existing trading relationships. This analysis and this example suggests that the investigation of network stability would be another important predictor of the likelihood of observing organizational births and the survival chances of already existing organizations and that the demographic (age-related) characteristics of such networks are critical.

If network stability is important, then factors that affect such stability are important foci of investigation. At the intraorganizational level of analysis, the argument has been made that the demography affects integration and cohesion (Wagner, Pfeffer, and O'Reilly 1984). In particular, individuals who enter the organization at the same time have fewer relationships already developed and yet need to develop networks for both task-related and social functions. Consequently, they are likely to develop interactions with each other, and interaction is argued to follow cohort lines. Extension of these arguments to the organizational level of analysis is straightforward. It suggests in the first place that firms that are of about the same age are more likely to trade with each other, ceteris paribus, and that transactions in markets may follow cohort lines where the cohorts are defined by the age of firms or by their date of entry into the market. It also suggests that entry into markets will be easier to the extent that there is more heterogeneity or diversity in the age of market participants. When there are relatively few and distinct cohorts, breaking into existing trading relationships, which are more likely to be more integrated and cohesive, may be more difficult. And, in general, the line of argument suggests that the demography of firms may be an important factor affecting the operation of markets and the structure of competition and entry, with these effects being particularly pronounced to the extent that transactions are importantly mediated by personal or institutional relationships.

Two other implications for analysis of the content of strategy also emerge fairly directly from this line of argument. One has to do with cross-cultural comparisons and, particularly, with the issue of understanding Japanese business and their comparative success. Although the role of the Ministry of Trade and Industry in actually organizing Japanese industry is debated, with some thinking there is a Japan, Inc., and others noting the tendency for firms to act independently of MITI's wishes, there is little doubt that Japanese industry is structured in a way that U.S. industry is not (see Clark 1979). There are stable customer-supplier relationships, not only in the automobile industry but in many others, as well as relationships among banks and industrial firms. Indeed, Clark (1979) has argued that Japanese firms are, on the whole, substantially less diversified (and often less integrated) than their U.S. counterparts but achieve many of the effects of diversification and integration through the development of stable and substantial interfirm organizations, bound together not only through commercial or trading relations and the transfer of capital but also frequently through the movement of employees from the central firm or firms to periphery organizations in times of economic slack and or at the time of retirement.

If we move the level of analysis up from industry sectors to nation states, this may help to explain Japanese industry success. For what may be occurring is an organized, fairly centralized sector trading with a diversified set of relatively unorganized sectors, exactly the conditions that should lead to structural autonomy and profit. The Japanese example suggests other cross-industry comparisons both within the United States and within Japan, as well as cross-national comparisons of the organization of industry as a predictor of profit and other outcomes. We clearly need much better information about the consequences of various forms of industry organization and structure, and here I refer to something much more than the level of concentration or the form of competition.

The second implication is that we also need to understand much better how industries do or do not come to be organized and structured. This might be accomplished in part by doing historical studies of industries. For instance, MacAvoy (1965) has written a very interesting book on the development of the Interstate Commerce Commission (ICC). He found that the railroads in the midwest were continually attempting to develop cartels in the mid-1800s to control the price charged for shipping grain to the east coast ports. The problems of cartel maintenance were severe because the industry had very high fixed and low variable costs and essentially sold an undifferentiated service, rail transportation. Thus, there were tremendous inducements for firms to cheat on the established cartel price, thereby generating additional revenues in the short run that exceeded their variable costs of providing the additional cargo carrying service. The history of the industry prior to the ICC was one of the cartel forming, breaking apart, and reforming. Of course, the creation of the ICC provided what the railroads had long needed, an effective sanctioning mechanism, and MacAvoy noted that once the ICC was created, freight rates went up to their highest and most stable level. A history of the railroads' involvement in the agency's creation, and as important, their attempts to influence its operation once created, would be an important contribution to our understanding of how industries organize. Similar studies might be conducted of other governmental economic regulatory agencies such as the Civil Aeronautics Board, utility rate regulating organizations, the Federal Communications Commission, and so forth. In this analysis, we need to go beyond the already numerous studies of the effects of such agencies on price, entry, and profits and to understand better how and why they were created and the specific strategic actions taken by the firms in the industry during this creation process and afterwards.

In addition to such natural histories, it is clear we need a much better understanding of how various interfirm arrangements actually affect centralization and coordination within sectors. As Burt has argued with respect to interlocking directorates, the presence of ties neither demonstrates cooptive intent by itself nor, certainly, cooptive effect. In this regard, the research that emanated from the resource dependence perspective on interfirm movement of executives (Pfeffer and Leblebici 1973), joint ventures (Pfeffer and Nowak 1976), and board of directors interlocks (Pfeffer and Salancik 1978) represented a very partial beginning of this line of inquiry. Demonstrating correlates of interfirm arrangements is not sufficient at all to indicate their effects; moreover, much more dynamic, time-dependent analyses of these processes are required. This area of inquiry remains an important and comparatively unexplored one for understanding how industries do or do not get organized.

There are also some implications of this recognition of the inter-dependent, relational nature of interorganizational activity to public policy. In particular, much of the current writing on strategy, such as Lawrence and Dyer (1983), Hayes and Wheelwright (1984), and Ouchi (1984) assumes a virtually complete correspondence of interests between private firms and public policy. The Panglossian argument is made that what is good for restoring profit and health to U.S. industries is also inevitably and invariably in the best interests of public policy and the U.S. consumer more generally. Such arguments are plausible only because these authors have either restricted themselves to internally focused responses that increase efficiency (such as manufacturing efficiencies as in Hayes and Wheelwright or organizational efficiencies as in Lawrence and Dyer) or because, as in Ouchi's case, the assumption is made that the inevitable conflicts of interest between organized industries among themselves and with other sectors of the society are best resolved when they are made explicit and worked out between formally organized and recognized contending groups. The first limited focus is clearly at best incomplete if not incorrect, and the second assertion that helping industries to organize themselves will serve the public interest is certainly an empirical question.

Rather, it is clear that profit is achieved in many ways and that its role as a signal of economic efficiency is not quite as good and pure as economic theory suggests. Profit is achieved by restricting imports, by obtaining industry subsidies from the government, by achieving the ability to legally fix price and restrict entry, and by more generally achieving power with respect to one's environment. Profit, measured in any way, is more than just the result of managerial good behavior and efficient production of what the customer wants. It is sometimes that, but sometimes not.

Most of these profit-enhancing actions come through political mechanisms, sometimes through bureaucratic agencies, sometimes through legislation, at various levels of government. There have been useful studies of the tobacco industry (such as Miles 1982; Grefe 1981) that indicate how the surgeon general's report discouraged new entry into the industry, which was also forestalled by heavy production and advertising barriers, thereby protecting cigarette makers' markets even as the manufacturers were able to achieve public actions that encouraged and facilitated cigarette export and forestalled action that delimited opportunities to smoke. The result is that most of even the most diversified cigarette companies' profits still come from the production and sale of cigarettes, not from the soft drinks, beer, or shipping lines and fast foods they diversified into.

This suggests that the role of the state and the production of public policy is inextricably linked to the study of strategy and of organizations, in a way that the current literature does not recognize at all. It might be noted that currently popular theories of public policy formulation retain much of the emphasis on methodological individualism and see governmental agencies as mere arenas in which the preferences and demands of individuals and other social actors become articulated and resolved. March and Olsen (1984) have recently critiqued this view of public institutions as mere settings, arguing for much more explicit attention to institutional factors in political life and the role of state institutions in both shaping as well as reflecting public attitudes and public preferences. Suffice it to say that the relationships of economic organizations to the state, broadly conceived, and a better understanding of the institutional factors of political life are requisite for the further development of a theory or practice of strategic management.

Finally, we can speak to the implications of this perspective for management, but here management is, as I alluded to earlier, well ahead of academic research. One set of implications has to do with the skills required by particularly those managers concerned with the formulation and implementation of strategy and the career paths by which such skills get developed and tested. If the goal of strategic management is the development of interorganizational power, then political skills, broadly defined, become critical. It becomes necessary to have skill at identifying and building coalitions of support, of being able to organize and mobilize sometimes diverse interests, of being able to effectively use political language (Edelman 1964), and of being sensitive to external, collective factors that affect the organization's well-being. There is some evidence that corporations are already beginning to recognize this through formal job rotation and career development policies that bring high-potential managers to the firms' Washington offices or provide them exposure and experience in trade association or other broader, interorganizational roles. We would hypothesize that such career paths, including hiring executives directly from government service, would be more important and more observed in industries needing to develop more structural autonomy and would furthermore be associated with enhanced performance to the extent such strategies of career development were successfully implemented. No industry perhaps better exemplifies this approach than the oil industry, which has been a model both of joint action and, in the present instance, of ensuring that executives on their way to the top achieved government and industry association exposure and experience. There are numerous possibilities for studies of executive career histories and how these have varied both over time, over industries, and even cross-nationally with respect to the incorporation of political or at least collaborative experience in the training and development profile.

Yet another implication is simply on the focus of attention of top management. If success is, at least partly, the consequence of the development and exercise of interorganizational power, and such power comes from the organization of one's own sector and the pattern of transactions and organization of other sectors, this suggests a focus on the task of strategic management that includes but also extends well beyond a fixation with product market entry and abandonment and the latest technique to try to wring more efficiency from the existing technology and work-force. It is indeed possible that in our focus on just-in-time inventory systems and lifetime employment and Type Z organizations (Ouchi 1981) we may have missed the critical, structural elements that contribute to the functioning, both good and not so good, of the Japanese economy. Here, too, the internal focus needs to be broadened for purposes of both academic research and policy analysis.

Finally, the skills of the strategic manager extend beyond those of implementing strategy within the organization to those involved in cross-organizational implementation. It is not by accident that when Peter Ueberroth stepped down as chairman of the U.S. Olympic Committee he had numerous offers from industry as well as from organized baseball. I believe there is growing, even if only implicit, awareness in industry that cross-organizational implementation of strategy, and the development of collaborative relationships across organizations is important and will become increasingly so. This is an area of research and practice that has yet to receive virtually any attention.

It is clear that the alternative conceptualization of business strategy discussed here broadens our view of strategic skills, strategic research issues, and what is involved in strategic management. In particular, the focus is directed outside the organization, to the structural and relational elements of societies and economies, and to the fundamentally political (in the broadest sense) nature of organizational strategy.

Conclusion

The argument made in this chapter is a simple one. As Presthus (1978) and Coleman (1974) have noted, we live in a society of organizations. But, as in most if not all social systems, that society is structured and furthermore is structured at times by institutions and by institutional arrangements such as governments, associations, professional associations, regulatory agencies, legislatures, and so forth (see Meyer and Scott 1983). The fact that relationships are patterned and that these patterns are at once social creations as well as determinants of social processes suggests that research on strategy incorporate this relational, institutional element to an extent it has yet to do. This requires moving away from the focus and emphasis on amorphous, undifferentiated environmental circumstances, broadening attention to incorporate a wider range of strategic actions and responses, and moving concern from internal adjustments and responses to attempts to manage, structure, and in other ways create a negotiated environment or order. It is a call to take relationships, quasi-firms, trade associations, and interfirm organizations of all types and other manifestations of networks, resource dependencies, and relations more seriously. Managers, it would seem, already are ahead of academic research in this regard, though theory is already available that can be used to help guide both research and practice of strategy in some new directions.

REFERENCES

Astley, W. Graham. 1984. "Toward an Appreciation of Collective Strategy." *Academy of Management Review* 9: 526–35.

Astley, W. Graham, and Charles J. Fombrun. 1983. "Collective Strategy: Social Ecology of Organizational Environments. *Academy of Management Review* 8: 576–87.

Baysinger, B.D., and W.H. Mobley. 1983. "Employee Turnover: Individual and Organizational Analysis." In *Research in Personnel and Human Resources Management, Vol. 1*, edited by K. Rowland and G. Ferris, pp. 269–320. Greenwich, Conn.: JAI Press.

Bower, Joseph L. 1970. *Managing the Resource Allocation Process*. Boston: Harvard Business School.

Burt, Ronald S. 1980. "Autonomy in a Social Topology." *American Journal of Sociology* 85: 892–925.

———. 1982. *Toward a Structural Theory of Action: Network Models of Social Structure, Perception, and Action*. New York: Academic Press.

———. 1983. *Corporate Profits and Cooptation: Networks of Market Constraints and Directorate Ties in the American Economy*. New York: Academic Press.

Caves, Richard E. 1970. "Uncertainty, Market Structure and Performance: Galbraith as Conventional Wisdom." In *Industrial Organization and Economic Development*, edited by J.W. Markham and G.F. Papanek, pp. 283–302. Boston: Houghton Mifflin.

Clark, Rodney. 1979. *The Japanese Company*. New Haven, Conn.: Yale University Press.

Coleman, James S. 1974. *Power and the Structure of Society*. New York: Norton.

Collins, Norman R., and Lee E. Preston. 1968. *Concentration and Price-Cost Margins in Manufacturing Industries*. Berkeley: University of California Press.

Eccles, Robert G. 1981. "The Quasifirm in the Construction Industry." *Journal of Economic Behavior and Organization* 2: 335–57.

Edelman, Murray. 1964. *The Symbolic Uses of Politics*. Urbana, Ill.: University of Illinois Press.

Freeman, John, and Michael T. Hannan. 1983. "Niche Width and the Dynamics of Organizational Populations." *American Journal of Sociology* 88: 1116–45.

Granovetter, Mark. 1985. "Economic Action and Social Structure: A Theory of Embeddedness." *American Journal of Sociology* 91: 481–510.

Grefe, Edward. 1981. *Fighting to Win: Business Political Power.* New York: Harcourt Brace Jovanovich.

Hall, William K. 1984. "Global Competition in Basic Industries: Some Predictions on the Next Round." Presentation at Graduate School of Business, Stanford University, October 17, 1984.

Hannan, Michael T., and John H. Freeman. 1977. "The Population Ecology of Organizations." *American Journal of Sociology* 82: 929–64.

Hayes, Robert H., and Steven C. Wheelwright. 1984. *Restoring Our Competitive Edge: Competing Through Manufacturing.* New York: John Wiley.

Hirsch, Paul M. 1975. "Organizational Effectiveness and the Institutional Environment." *Administrative Science Quarterly* 20: 327–44.

Hrebiniak, Lawrence G., and William F. Joyce. 1984. *Implementing Strategy.* New York: Macmillan.

Lawrence, Paul R., and Davis Dyer. 1983. *Renewing American Industry.* New York: Free Press.

MacAvoy, Paul W. 1965. *The Economic Effects of Regulation.* Cambridge, Mass.: MIT Press.

Mackenzie, Kenneth D., and George D. Frazier. 1966. "Applying a Model of Organization Structure to the Analysis of a Wood Products Market." *Management Science* 12: B-340–52.

March, James G., and Johan P. Olsen. 1984. "The New Institutionalism: Organizational Factors in Political Life." *American Political Science Review* 78: 734–49.

Meyer, John W., and W. Richard Scott. 1983. *Organizational Environments: Ritual and Rationality.* Beverly Hills, Calif.: Sage.

Miles, Robert H. 1982. *Coffin Nails and Corporate Strategies.* Englewood Cliffs, N.J.: Prentice-Hall.

Mumford, Enid, and Andrew Pettigrew. 1975. *Implementing Strategic Decisions.* London: Longman Group.

Ouchi, William. 1981. *Theory Z.* Reading, Mass.: Addison-Wesley.

———. 1984. *The M-Form Society.* Reading, Mass.: Addison-Wesley.

Pfeffer, Jeffrey. 1972. "Merger as a Response to Organizational Interdependence." *Administrative Science Quarterly* 17: 382–94.

Pfeffer, Jeffrey, and Huseyin Leblebici. 1973. "Executive Recruitment and the Development of Interfirm Organizations." *Administrative Science Quarterly.* 18: 449–61.

Pfeffer, Jeffrey, and Phillip Nowak. 1976. "Joint Ventures and Interorganizational Interdependence." *Administrative Science Quarterly.* 21: 398–418.

Pfeffer, Jeffrey, and Gerald R. Salancik. 1978. *The External Control of Organizations: A Resource Dependence Perspective.* New York: Harper & Row.

Phillips, Almarin. 1960. "A Theory of Interfirm Organization." *Quarterly Journal of Economics* 74: 602–13.

Porter, Michael E. 1980. *Competitive Strategy: Techniques for Analyzing Industries and Competitors.* New York: Free Press.

Presthus, Robert. 1978. *The Organizational Society.* New York: St. Martin's Press.

Wagner, W. Gary, Jeffrey Pfeffer, and Charles A. O'Reilly III. 1984. "Organizational Demography and Turnover in Top Management Groups." *Administrative Science Quarterly* 29: 74–92.

Weiss, Leonard W. 1963. "Average Concentration Ratios and Industrial Performance." *Journal of Industrial Economics* 11: 237–54.

7 THEORY, STRATEGY, AND ENTREPRENEURSHIP

Richard P. Rumelt

Where do new businesses come from? The textbooks say that the entrepreneur, like the stork, brings them. But new businesses do not occur with equal likelihood in all societies or all industries. Also, existing firms in advanced societies have finely developed methods for managing diversified portfolios of businesses, so it is unclear why so many risky *new* businesses are formed. Why don't existing firms, with their experience, established reputations, and in-place resources, have compelling advantages in new business formation? This chapter examines the locus of entrepreneurship, both in terms of product-market conditions and organizational context.

Schumpeter (1950) described the entrepreneur as combining resources in new ways. In this vein I define *entrepreneurship* as the creation of *new* businesses, and by *new* I mean businesses that do not exactly duplicate existing businesses but have some element of novelty. For example, the entrepreneur may be opening a convenience store in a hitherto untried location, may have developed a new product or a new production technology, may have a new way of promoting a product, may have identified a novel market segment, or may be betting on a novel method of distribution. I do not automatically equate entrepreneurship with the creation of new organizations or ventures, although I will be concerned with the conditions impeding internal entrepreneurship.

If entrepreneurial activity is seen as motivated by the chance for gain, its frequency, locus, and organizational context should be determined by the availability of entrepreneurial insights, by the potential returns to entrepreneurship, and by the entrepreneur's ability to attract the requisite resources. A good working theory of entrepreneurship would begin with these principles and develop connections to observable and predictable phenomena. It would be useful, for example, to be able to characterize the systematic differences in the potential for entrepreneurial gain across product groups, industries, and societies. In addition, it would be good to have more precise understanding of the types of structural and contractual arrangements that facilitate or impede entrepreneurial activity.

This chapter explores the terrain on which theories of entrepreneurial activity might be built. In the next section I examine the product-market context of entrepreneurial activity, focusing on the availability of *entrepreneurial rent* and the conditions enhancing its availability and inhibiting its appropriation. The following section explores the organizational context of entrepreneurship, analyzing some of the factors favoring and inhibiting internal entrepreneurship. These ideas are then drawn together in a simple framework for predicting entrepreneurial activity.

The Product-Market Context of Entrepreneurial Activity

Since John Stuart Mill introduced the idea of the "stationary state," economists have tended to see the real world as a deviation from some ideal stable condition. Indeed, the central result of neoclassical microeconomics is that individual profit (or utility) maximization in a perfectly informed frictionless economy eliminates any resource waste and drives profits, though maximized, everywhere to zero. This model, however, has nothing to say about the source of new businesses, new products, innovations, or new ways of doing things. As Schumpeter emphasized, the competitive ideal not only fails to describe entrepreneurship, it fails to provide a motive for the search for new methods. If competition is swift and frictionless, entrepreneurs can expect only zero profits if projects succeed and worse if they fail!

The Industrial Economics Tradition

Because of the power and acceptance of the competitive model, the economic analysis of innovation and entrepreneurship has been only weakly concerned with the description of real events; instead, its focus has been the critique of the competitive model's descriptive or normative validity. Thus Schumpeter, the originator of the economics of innovation and entrepreneurship, argued that innovation was incompatible with the competitive ideal, since the risk and cost of innovation would not be voluntarily borne without the possibility of compensating gains. These gains, he stressed, appeared in the form of the high profits earned by monopolists and tight oligopolies. Eliminate monopoly power and you throttle innovation.

In the same spirit, Galbraith took the position that innovation was the province of large firms. He argued (1952: 91) that "most of the cheap and simple innovations have, to put it bluntly, already been made," so that only large firms earning monopoly profits could afford to undertake the costly search for new products and techniques.

A respectable literature has grown around the discussion and empirical testing of these ideas. Theoretical work has been pressed by Arrow (1962), Demsetz (1969), Nelson and Winter (1982), Kamien and Schwartz (1982), and others. Important empirical studies have been performed by Mansfield (1968, 1971), Scherer (1965, 1967), Comanor (1967), Phillips (1971), and Williamson (1965).[1] The approach to the issue that has evolved, especially in empirical work, has been to equate market structure (read *concentration*) or firm size with monopoly power and to examine the connection between monopoly power and innovation, the latter usually measured by R&D spending or patenting.

This work is interesting and useful, but its very volume should make it obvious that no clear-cut conclusions have emerged. The best that can be said in general is that innovation does not appear to be strictly the province of the large firm or of oligopolists. The problems with this literature, however, extend beyond its lack of plain answers. In the quest to clarify and test Schumpeter's and Galbraith's assertions, researchers have come to accept a number of questionable propositions. In particular, they have tended to (1) identify all rents as monopoly rents, (2) to equate firm size (or concentration) and market power, (3) to restrict the definition of innovation to technological invention, (4) to assume that R&D spending is the source of invention, and (5) to identify patents as the measure of invention.

That entrepreneurial innovation need not be technical should go without saying. The new form of package delivery service created by Federal Express was innovation, as was the CMA Account developed by Merrill Lynch and the development of pay cable TV channels. Drucker (1985: 31) reminds us that

> Innovation . . . does not have to be technical, does not indeed have to be a "thing" altogether. Few technical innovations can compete in terms of impact with such social innovations as the newspaper or insurance. Installment buying literally transforms economies. Wherever introduced, it changes the economy from supply-driven to demand-driven, regardless almost of the productive level of the economy.

The equating of the returns to entrepreneurship with monopoly power, and the subsidiary association between size, concentration, and monopoly, is a more fundamental problem with much research on innovation. First let us examine the term itself. What, exactly, is a *monopoly profit*? If all profits in excess of fully competitive returns are called *monopoly profit*, the term has no special meaning. It should be obvious that investments in risky entrepreneurial projects can be justified only if the losses on failure are balanced by above-normal returns associated with success. If, for example, totally specific capital is committed to a venture with a one-in-two chance of complete failure (loss of the investment), then the profit rate on success must be twice the normal rate (assuming annuities) to justify investment. Are such profits, if achieved, monopoly profits? In the static theory monopoly profits derive from the artificial restriction of competitors' outputs, are a distortion, and imply waste. That is, once the innovation has been accomplished, the excess profits could be appropriated without curtailing the supply of the new product or service. However, such policies would diminish the supply of innovation in the first place. So if we desire a theory wherein innovation is endogenous, it is incorrect to use the term *monopoly* with regard to entrepreneurial returns. The issue is not one of monopoly but the quite traditional problem of the proper allocation of property rights.

The equating of monopoly profit with size and concentration is also a problem in this stream of research. Put directly, the market power framework posits that firms earn surplus profits by colluding behind strategically erected entry barriers. (The entry barriers by themselves are not sufficient for without diminished competition those behind the barriers would erode each others' profits.) Yet innovation and entrepreneurship are really about novelty and differentiation; models of commodity-producing collectives may not be the best approach to their study. An alternative viewpoint, one that emphasizes the uniqueness of firms and identifies profits with resource bundles rather than with collectives, is offered by the strategy field.

The Competitive Strategy Tradition

The systematic study of business strategy, as practiced in schools of business and management, had its beginnings in case studies of several firms within an industry. These investigations revealed that firms in the same industry often differed markedly from one another. Although operating in the same basic competitive environment, the managements of different firms were seen to have adopted different policies regarding product quality, line breadth, distribution channels, financial leverage, and employee relationships, and they were observed to use different organizational structures. In addition, there were usually substantial and sustained differences in performance among the firms within an industry.

These differences among close competitors were identified as differences in *strategy*, and the field of study has concentrated on understanding strategy in both descriptive and normative terms.

The first basic theory that arose from these data was that of *fit*. According to this framework, a high-performing firm had a product-market strategy that was consonant with the opportunities and constraints imposed by its competitive environment and additionally had an organizational structure suited to its strategy. Good management consisted of the alert tracking of competitive conditions and the implementation of concomitant adjustments in strategy and structure.

The trouble with the fit theory is that it failed to adequately explain why all competitors were not fit. If fit leads to success, and firms are similarly motivated toward success, why are there unfit strategies? To adequately answer this question, strategy researchers have turned to concepts that emphasize the special histories and resource bundles of each firm. Caves and Porter (1977) see firms as having initially different "traits" and strategically moving to build competitive positions around these differences. Lippman and Rumelt (1982) model differences among firms as stochastically generated and as difficult to imitate because of causal ambiguity regarding their sources. Wernerfelt (1984) emphasizes the importance of unique resources (resource barriers) to business strategy. Hitt and Ireland (1985) explored the empirical association between firm distinctive competence and performance.

Empirical work also reveals that the dispersion of long-term profit rates within industries is very much larger than the dispersion of industry profit rates across industries. For example, applying a variance components analysis to rates of return on capital displayed by 1,292 U.S. corporations over a twenty-year period I obtained the results shown in Table 7–1.[2] The data show that the variance in long-run profitability *within* industries is three to five times larger than the variance *across* industries. Clearly, the important sources of excess (or subnormal) profitability in this data set were firm specific rather than the results of industry membership. Once the source of high profits is located in the firm's resource bundle

Table 7–1. Results of Variance Components Analysis of Return on Capital, 1,292 U.S. Corporations

	Industry Definition	
	3-digit	4-digit
Variance due to industry effects	3.9	4.7
Variance due to firm effects within industries	19.2	17.6

rather than in its membership in a collective, the appropriate profit concept is that of *rent*.

The Concept of Entrepreneurial Rent

The idea of economic rent was developed in about 1820 by David Ricardo, as part of his argument for the abolition of England's Corn Laws. Ricardo noted that land varied in fertility, so that when demand was sufficient to make it economic to grow corn on less fertile land, high profits were earned by anyone owning very fertile land. These extra profits were called *rents* because they ultimately accrued to the owners of the land. Some commentators argued (as in today's rent-control battles) that corn was expensive because of the large rents paid to land owners. The heart of Ricardo's (1971) argument was that the price of corn was determined by the supply of fertile land and not the level of rents:

> Corn is not high because a rent is paid, but a rent is paid because corn is high; and it has been justly observed that no reduction would take place in the price of corn, although the landlord should forego the whole of their rent.

Ricardian Rents. The differences in payments received by factors of the same "type" are Ricardian rents. The factors are, of course, not exactly of the same type else no rents would be paid. The key to the existence of Ricardian rents is the presence of a fixed scarce factor; the scarcity is such that the extra profit (rent) commanded by this factor is insufficient to attract new resources into use. A standard way of presenting this notion is the *increasing cost* industry. In this type of industry, it is possible (at some given price) to rank the producers from least cost to highest cost, with the marginal cost of the least efficient producer equal to the market price. The marginal firm earns zero profit while the more efficient firms earn rents. The surplus profits in this case (assuming atomism) are not socially objectionable because the profitable firms' outputs are constrained by fixed factors rather than restricted as a stratagem to raise the market price.

The rent concept due to Pareto (and Marshall) is the difference between a resource's payments in its best use and the payments it would receive in its next best use. Thus, the *Pareto rent* is the payment received above and beyond that amount required to call it into use. When resources in use all have the same value in their best alternative use, the Ricardian and Pareto concepts correspond.

Rents, unlike *profits*, persist in static equilibrium. The usual microeconomic treatment of rents is to ascribe them fully to the scarce factor and then to treat that factor as separately owned, so that the firm's costs include the rent. If the scarce factor is then traded, the rents are capitalized and no one (except some original owner) shows any profit. This formulation is traditional and saves the zero-profit condition of neoclassical theory. It is inadequate, however, in the face of newer insights. In particular, we now understand that resources that can just as well be rented as owned are of a very special type: They are nonspecific and their use can be obtained via market mechanisms with minimal transaction costs. If, however, the fixed rent-yielding factor is specialized to the needs of the firm, or if its use otherwise involves significant transaction costs, the rent on that factor is not logically or operationally separable from the profits of the firm.

Entrepreneurial Rents. The classical concept of rent applies in a static world and compares the productivity of different resources or of a resource in different uses. Entrepreneurship, by contrast, is the discovery of new combinations of resources and uncertainty is the central issue. I therefore define *entrepreneurial rent* as the difference between a venture's *ex post* value[3] (or payment stream) and the *ex ante* cost (or value) of the resources combined to form the venture. If we posit expectational equilibrium (*ex ante* cost equals expected *ex post* value), then expected entrepreneurial rents are zero. The basic thrust of this definition is to identify those elements of profit that are the result of *ex ante* uncertainty.

Although rents are not competed away in normal competition, they can be appropriated because they are payments for a factor above and beyond that required to attract it to its present use.[4] Thus, if a restaurant is yielding $500,000 per year in profit but would have recovered all the costs of planning, capital, and set-up if it earned profits of only $300,000, the difference, $200,000, is rent. The rent is appropriable in that one could reduce the restaurant's profits by $200,000 (keeping prices the same) without seeing it reduce its level of operations. The *ex post* appropriability of entrepreneurial rent means that owners of rent-yielding assets must anticipate the erosion of rents as interested individuals, groups, and governments opportunistically seek to redefine their *shares*. In addition, entrepreneurial investments are necessarily specialized to a specific (novel) use, or else there would be no risk of loss. Therefore, the entrepreneur also faces the possibility of appropriation of the additional rents accruing to the specialized portion of the original investment.

Interestingly, the rent-earning firm looks much like the classical successful enterprise of the strategy literature:

It exhibits a high profit rate and substantial discretion in the allocation of its profit stream.
At its core rest unique specialized resources that cannot be freely expanded or imitated.

Its management perceives it as vulnerable to the political bargaining and legal maneuverings of unions, governments, consumer groups, and so forth.

Uncertainty and Rent

Given expectational equilibrium, it is uncertainty that produces the possibility of entrepreneurial rents. Absent uncertainty, we would expect the inputs used in the entrepreneurial venture to reflect their value in use or we would expect *ex ante* crowding or rapid imitation to reduce profits to normal levels. This uncertainty is normally viewed as *discovery* or invention. The two basic kinds of entrepreneurial discovery concern the value of resource combinations and the pattern of demand.

The entrepreneurial discovery of resource value includes mineral exploration, real estate development, technological invention, and the creation of new means of producing and delivering products and services. The discovery of demand patterns includes satisfying new consumer needs and wants and identifying new market segments worthy of attention and focus. Where entrepreneurial activities completely resolve the original uncertainty, the results achieved, absent secrecy, could be perfectly imitated. In this case it is best to provide the innovator with property rights that encourage the dissemination of knowledge. If however, the venture leaves considerable residual uncertainty, as is often the case in commercial rather than technical innovation, the entrepreneur faces a moral hazard problem in obtaining payments from others for what has been learned.

In the limiting case of Lippman and Rumelt's (1982) "uncertain imitability," the causal ambiguity is so great that successful entrepreneurs are no more likely to repeat their success than *de novo* entrants. Here information dissemination is valueless and consequently cannot be a source of entrepreneurial return.

Rent Size and Durability

What permits a risky entrepreneurial venture to earn rents if it succeeds? The business must be a sufficient innovation to be an efficient replacement for substitutes, it must resist the appropriation of rents, and it must have some protection against imitative competition.

The first condition is simply that the innovation be socially efficient. That is, it must provide a sufficient increment in value over pre-existing substitute products or technologies to justify the costs of innovation. Where such gains are not possible, entrepreneurial innovation cannot begin to pay for itself.

The primary appropriation challenges entrepreneurs face are those due to powerful buyers or suppliers (including employee groups), the owners of cospecialized assets, and governments. If the venture uses inputs from a monopolist, or sells its output to a monopsonist, it faces a complex bilateral bargaining situation. Even if contracts have been hammered out before the venture is complete, the powerful buyers or suppliers have incentives to opportunistically recontract, raising the costs of the venture or reducing its returns. A special type of supplier problem occurs when the entrepreneur needs the services of a cospecialized asset. For example, an innovator who develops a new household cleanser would face the prospect of choosing between building a new sales and distribution system or bargaining with a giant household products firm to obtain distribution services. Teece (forthcoming) provides a useful discussion of the contracting options open to such an entrepreneur.

Isolating Mechanisms. Given an innovation expected to be socially efficient, and absent appropriation challenges, entrepreneurship will not be justified unless there are impediments to the immediate *ex post* imitative dissipation of entrepreneurial rents. I call such impediments *isolating mechanisms* (Rumelt 1984) in rough analogy to the ecologist's use of the term to describe barriers to species mobility.

Among the most important isolating mechanisms are property rights. In the early days of the oil industry, for example, the Rule of Capture defined oil as a migratory good (like fish or wild game) and assigned possession only to those who extracted it from the ground. This assignment of property rights, together with the fact of multiple leases on each reservoir, led to very rapid exploitation of new oil fields. Overpumping depressed market prices, which, in turn, reduced incentives to search for oil. A better assignment of property rights would have prevented wasteful overdrilling in known reservoirs and underexploration for new ones. Similarly, fewer resources will be devoted to the quest for an invention that is easily imitated than for one of equivalent efficiency but that can qualify for effective patent protection.

Although the law provides the entrepreneur with property rights over discoveries of minerals, patentable inventions, written material, and trademarks, no such protection exists for the vast bulk of business innovation. New packaging concepts, methods of distribution, manufacturing methods and planning techniques, consumer research methods and information, and most new product ideas entail no assignment of property rights. Were imitative competition in these areas immediate and perfectly frictionless, none of these innovations would be sought. Fortunately, there are numerous lags, information asymmetries, and frictions that function as *quasi-rights*, thereby sustaining entrepreneurial rents.

The isolating mechanisms that protect entrepreneurial rents from imitative competition normally appear as *first-mover advantages*. That is, they are asymmetries, usually derived from informational inequalities or the costs of creating and enforcing complex multiparty contingent contracts, that, other things equal, make it increasingly costly for followers to duplicate an innovator's position. There is no unambiguous mutually exclusive list of these phenomena, but the most important appear to be as follows:

Information impactedness: When innovators can prevent potential competitors from obtaining the knowledge gained from successful operation of a venture, they can inhibit effective imitation. Secrecy is obviously more difficult where the knowledge is scientific rather than tacit, where more people are privy to the information, and where employee mobility is high. In the limit, where uncertain imitability holds, competitors cannot extract the innovator's secrets because even the innovator does not know the causes of success.

Response lags: Competitors may be slow in responding to an innovator, providing high entrepreneurial rents in the interim. Such lags may be due to the time it takes for competitors to recognize, evaluate, and formulate a response to the innovation, or may simply be due to waiting times for specialized equipment. Lags also occur because competitors are unwilling to cannibalize existing high-rent businesses or because of legal constraints. For example, on deregulation, AT&T was prevented by law from meeting MCI's prices on long-distance voice communications services for a period of seven years.

Economies of scale: If the minimum efficient scale of a business is comparable to the size of the market, and if the assets required are specialized to this use, a traditional *entry barrier* occurs. Additional entry is deterred by the prospective entrants' recognition that adding another efficiently sized competitor to the business would depress price below full cost.

Producer learning: In certain cases a producer becomes more efficient as experience is gained, measured by the passage of time or by cumulative output. If the knowledge base underlying this efficiency gain is tacit, so that it resists transfer to other producers, competitors with less experience are at a comparative disadvantage. Producer learning appears to be most important in operations where complex assembly operations are performed.

Buyer switching costs: If early buyers of a new product find it subsequently costly to switch to a competitor's offering, the first mover is at an advantage. Buyer switching costs are high when the product is durable and specialized, when there are substantial specialized co-investments that the buyer must make, where search or evaluation costs are high, or where buyers invest substantial specialized human capital in learning how to use or consume the product. Even though a follower's product is technically superior to the innovator's, buyer switching costs may prevent its adoption. The problem is technically one of contracting costs: If the buyers could costlessly enter into a mutual contract to wait for the follower's better product, they could diminish the innovator's profits and better themselves.

Reputation: Many products cannot be accurately evaluated by buyers until after they have been purchased and used. As Klein and Leffler (1981) show, a producer's ability to sell high-quality versions of such *experience goods* depends on its *reputation.* To the extent that buyers' beliefs about reputation depend on the length of time the producer has operated reputably, first movers can obtain reputational advantages. Of course, other things may not be equal, and the innovator may face imitators who have substantial reputations built up over time in related businesses (for example, Apple versus IBM in personal computers).

Communication good effects: Certain products increase in value as the number of adopters or users increases. Examples are telephone network services, microcomputer software, and audio compact disk players. Connor and Rumelt (1986) term these *communication goods.* The effect arises because the product serves as a means of social coordination (standardization) or because a larger user base calls into being a larger number of complementary goods. When communication goods are also experience goods (such as microcomputer spreadsheet software), there is a market need for both standardization and reputation-bonding. The upshot is the *de facto* standard, where a particular brand or manufacturer's product becomes the means of coordination. These competitive positions are *very* powerful and offer the promise of large entrepreneurial rents.

Buyer evaluation costs: As buyers face increasing problems in evaluating competing products they seek ways of economizing on evaluation costs. The most common tactic is to free-ride on the presumed analyses of the well informed and to buy the market leader. Such behavior provides advantages to the market leader as long as the follower's product is not significantly better.

Advertising and channel crowding: Early entrants into a market sometimes face less crowded advertising message spaces and distribution channels. When the first compact personal low-cost plain paper copying machines appeared, for example, Canon's advertisements stood out sharply because no other manufacturer offered a comparable product. Several years later, as the fifth manufacturer attempts to enter the market, it is much more difficult to get the buyer's attention. The multiplicity of similar messages dims the impact of all. This asymmetry allows the early entrant to build customer awareness less expensively than later entrants. A similar effect occurs with distribution channels. Distributors and retailers face fixed set-up costs associated with taking on new lines (billing systems, salesperson training, and so forth) and minimum fixed costs associated with handling a line of products (allocation of shelf space, spare parts supplies management, and so forth). Consequently, there is room in distribution channels for only a limited number of essentially similar product lines. Late entrants into a market must either chase niche segments or buy distribution by paying substantially larger dealer margins.

The Product-Market Locus of Innovation

The amount of society's resources devoted to entrepreneurship will depend on *ex ante* estimates of entrepreneurial rents and the level of uncertainty. As the potential size of entrepreneurial rents increases, the prizes get larger and more entrepreneurial activity can be expected.

Given limited liability and the right to cease operations and break contracts through declarations of bankruptcy, it is very possible that entrepreneurial activity will *increase* with increases in uncertainty. That is, if the chances of very positive outcomes are increased, and the losses due to negative outcomes are limited, then more uncertainty can lead to a larger expected value of innovation.[5]

The idea that entrepreneurship increases with uncertainty probably explains the common perception that entrepreneurs are risk-takers. For example, in Grayson's (1960) classic study of oil and gas operators' drilling decisions, his assessed utility functions on wealth were convex, implying risk-seeking attitudes. But it is very possible that these operators had difficulty separating their attitudes toward risk *per se* from their perceptions about the values of various ventures. In equilibrium, ventures with higher uncertainty (holding the mean constant) about the amount of oil below ground are worth *more*. It is likely, therefore, that Grayson's data reveal the wildcatters' preference for increases in uncertainty over the size of the find rather than for financial risk.

The factors influencing the size and duration of entrepreneurial rents will also have a marked effect on innovative activity. Obviously, where appropriation is common, through either government action or opportunistic bargaining by powerful parties, entrepreneurship is reduced. In addition, it is clear that projects involving important cospecialized assets will have the largest expected yield to the owners of those assets, placing the probable locus of entrepreneurship within existing organizations in such cases.

Finally, it is useful to note that much of the initial uncertainty attached to a really novel entrepreneurial venture concerns the strength and quality of the isolating mechanisms that will be present. When RCA undertook its venture in videodiscs for home entertainment, there was uncertainty concerning consumer response. There was also great uncertainty as to the size of any first-mover advantages that might accrue and as to the ability of film companies to eventually appropriate the profits. The venture's failure resolved the consumer response question but left the issue of appropriability and isolating mechanisms open.

In another example, early entrepreneurs in the microcomputer software industry expected that publishers would be distributing a wide variety of titles to the public, envisioning thousands of competing titles. They were taken by surprise when early products (such as dBase II and WordStar) became huge bestsellers and proved difficult to displace even by superior products. As the *de facto* standard aspect of the microcomputer software industry became apparent (reputations plus communication effects), a large increase in entrepreneurial effort followed this increase in expected gross entrepreneurial rent. This industry exhibited large rents for the first movers, but their very staying power naturally leads to diminished entrepreneurial effort once it is perceived that the key niches have been filled.

In many industries, after the first wave of innovation, competition is aimed at reductions in the size of isolating mechanisms. Thus, if buyer learning is an important advantage for first movers, easier to learn products may be developed. If producer learning is crucial, more automated process-like methods will be tried by those seeking to undermine the leader's experience. If channel crowding is the source of advantage, followers will seek out new forms of distribution. These competitive moves, themselves innovative activity, all act to carry the industry from its early birth stages to maturity. As the industry matures, early entrants must try to understand whether the industry will become rent-free or whether it will contain protected niches for those who play correctly.

The Organizational Context of Entrepreneurial Activity

Given the product-market conditions for entrepreneurship, which organizations will innovate and when will innovation be carried out within new ventures rather than in existing firms? I will first look at the total organizational incentives to innovate, treating the firm as a single actor, and then I examine the problems of entrepreneurship from the perspective of the individual member of the firm.

The Problem of Cannibalism

It was a commonplace in Detroit during the 1950s that small cars were less profitable than larger cars and that the wise manufacturer did not cannibalize a profitable midsized auto business by promoting less expensive small cars. Similarly, it can be argued that Xerox's incentive to respond to low-price Japanese plain-paper copiers was dulled by the possible cannibalization of its profitable higher volume machines. Jacobson and Hillkirk (1986: 15) note that

> The low-volume market is a low-margin business. The high-volume market . . . has always been a high-margin business.
>
> Of course, Xerox is afraid that low-volume products—whether Japanese- or Xerox-made—will pull business away from the crucial high-volume, high-margin end of the business.

Economists studying this issue have formulated the problem in terms of an incumbent monopolist deciding how hard to work on the development of a more efficient but lower-profit substitute.[6] The incumbent would just as soon never see the substitute appear, but others are also working on developing the substitute. Because the incumbent's gain from innovation is reduced by the destruction of the rent stream attached to the old product, the incumbent has less incentive to innovate and therefore spends less, at the margin, on innovative activity. The interesting thing about this insight is that the larger the original rent stream, the lower the incumbent's incentive to innovate.

Unlike the economist's model, the examples just cited identify the businessperson's concern with response to existing rather than potential competition. Were the incumbent's and rival's product perfect substitutes, there would be no reason for hesitancy; if the incumbent does not make and sell the new product, the rival will. But in many situations there are crucial asymmetries in customer response. In particular, customers may have established relationships with a vendor. They may have invested in learning about a vendor's product, they may depend on vendor-specific cospecialized services (such as service, brokers, dealerships), or they may depend on the vendor for tidings about new product events. When such customer relationships exist, it is reasonable to expect these customers to respond more positively to the vendor's introduction of a new substitute product than they would to a similar introduction by a competitor.

It is this differential response that produces the cannibalism problem. In Xerox's case, the company probably expected their traditional lease customers to respond more aggressively to a new line of low-cost Xerox copiers, with the consequent returns of on-lease midprice machines, than they would to the Japanese vendors' products. AT&T presently faces a similar problem with respect to the millions of telephones it has leased to the public. Rented at rates corresponding to purchase prices of $100 and more, this lease base provides the firm with enormous cash flow and dramatically curtails its incentive to aggressively compete in the new low-cost ($25) telephone business.

The cannibalism effect implies that in many cases the rent-earning incumbent will not be the innovator. Alternatively, it can be seen that the most fruitful approach for an entrepreneur may well be a direct attack on a profitable incumbent—such a firm may be least willing respond to the attack.

Organizational Routine

There is a vast literature on the issue of bureaucracy and the difficulty of obtaining change within large complex organizations. The issue can be framed in terms of bounded rationality, collective choice, or politics. Crozier (1964: 225) put it this way:

> People on top theoretically have a great deal of power and often much more power than they would have in other, more authoritarian societies. But these powers are not very useful, since people on top can act only in an impersonal way and can in no way interfere with the subordinate strata. They cannot, therefore, provide real leadership on a daily basis. If they want to introduce change, they must go through the long and difficult ordeal of a crisis. Thus, although they are all-powerful because they are at the apex of the whole centralized system, they are made so weak by the pattern of resistance of the different isolated strata that they can use their power only in truly exceptional circumstances.

There is also a life-cycle view of bureaucratic organization that holds that change becomes less possible as the organization ages. Downs (1967: 20) emphasized this aspect of bureaucracy, noting that "all organizations tend to become more conservative as they get older, unless they experience periods of very rapid growth or internal turnover."

Interestingly, there is also a large literature wherein the opposite is argued—that the large firm is the ideal environment for innovation. Shumpeter (1950), for example, claimed that the modern corporation had "routinized innovation," and Galbraith (1952) saw the resources and sustained collective action required for modern large-scale innovation as being most efficiently provided by large profitable firms. The weight of the empirical evidence on technological innovation does not show either economies or diseconomies of scale; no comparable work appears to have been done with regard to commercial and general nontechnical innovation.

Given the results of the technological innovation studies, there is no reason to suppose that large organizations are any less (or more) innovative than small or new organizations. What may be true is that the type of entrepreneurship differs. The best entrepreneurial opportunities for large organizations may be those based on the redeployment of the firm's resources and the extension of its competitive positions. Those most attractive to individuals and small firms may be based on new opportunity and the creation of new markets. For example, with the coming of airline deregulation, new entrepreneurial firms entered the industry with strategies based on non-union workforces and low-cost no-frills service. The established carriers, by contrast, worked to develop hubs, frequent-flyer plans, and created a whole new pricing technology for more effective price discrimination.

The Problem of Incentives

To many the essence of the entrepreneurial act is the acquisition of resources, but when the wealth at risk is not the entrepreneur's own, there is a potential problem of incentives. Arrow (1962) was the first to clearly define the problem as one of moral hazard. In his view, the separation of risk-bearing from innovation could be accomplished by simply paying the innovator a fee *as long as it is costless to monitor and evaluate the innovator's work.* But such control is not costless. Consequently, the innovator must be forced to bear at least some of the risk to ensure that he is actually delivering the agreed-on effort. Because the innovator may not have a taste for risk-bearing, too little innovation might be supplied in equilibrium.

A theoretical extension of this idea by Leland and Pyle (1977) shows that outsiders' valuation of an entrepreneurial venture depends on the proportion of the entrepreneur's wealth that has been placed at risk in the project. Downs and Heinkel (1982) provide some empirical support for the proposition that the value of investor's shares rises with the entrepreneur's personal commitment to the project.

These analyses are couched in market terms—they envision the entrepreneur as creating a new venture and having the problem of attracting investment funds. Does the problem of entrepreneurship within an existing firm have a similar structure? I will argue that the nature of the employment contract, managerial mobility, and less-than-perfect markets for managerial labor create incentive problems of a different kind. Rather than a reduction in innovation *per se*, there may be institutional myopia, wherein the organization's implicit discount rate on future income is higher than its cost of capital.

In organizations so large that decisionmaking is a multilevel process, analysis, proposal, and authorization are separate events. A number of researchers have observed that the authorization step is carried out in the face of large information asymmetries. Schon (1967: 110) observed that

> Entrepreneurs without authority cannot take the necessary leaps; their justifications before the fact always turn out to be inadequate. Both boss and subordinate operate in ignorance—one, in ignorance of the facts, opportunities, and problems of the innovative process; the other, in ignorance of the considerations which will be governing in making decisions.

In a similar vein, Mintzberg, Raisinghani, and Theoret (1976: 260) noted that

> In capital budgeting as well as in less formal types of authorization, a major problem is presented by the fact that the choices are made by people who often do not fully comprehend the proposals presented to them. Thus, in authorization the comparative ignorance of the manager is coupled with the inherent bias of the sponsor.

Given limited information, how is the authorization decision made? Bower (1967) studied the process in detail and argued that decisions are ultimately made on the basis of the proposing manager's track record. That is, by gradually building a reputation for reliable judgment, the lower-level manager gains credibility with senior management. The top-level managers cannot assess the projects *ex ante* but are somehow able to attribute reputations from assessments of managers' performance after the fact. One obvious problem with this administrative arrangement is that the top managers' ability to form accurate reputational estimates is severely limited by their presumed inability to comprehend the project *ex ante*. Additionally, the distribution of information leading to entrepreneurial projects will not necessarily correspond to the pattern of reputations. There is, by contrast, every reason to expect that younger managers with shorter track records will have fresher ideas and superior first-hand market and technological information.

Next, consider the impact of managerial mobility on decisionmaking in this context. Assuming that lower-level managers rationally attempt to maximize the net present value of their future earnings, how will managers behave? Given mobility, the manager must temper his view of how a project's future influences his reputation or income with the possibility that he will no longer be in the organization.[7] The net effect is that mobile managers will discount future cash flows more heavily than would be indicated by their personal discount rates on wealth or their employer's cost of capital. Given the fact that top management must choose among the projects that are actually proposed, the corporation as a whole will appear more myopic than are its members.

If managerial mobility is not just exogenous but potentially opportunistic, even more severe myopia can appear. Entrepreneurial managers, in competition with other managers for scarce project approvals, may sometimes find it necessary to misrepresent the future returns to a project. Calculating that they can leave the firm (or division) if it really begins to appear that their glowing promises will not be realized, these managers may select and support projects that show near-term gains but long-term losses. In essence, they hope to gain the reputational or pecuniary advantages associated with project acceptance and early returns and to avoid the penalties connected with future failure.[8] In part, they bet that their closeness to their projects will give them early warning, permitting opportunistic exit before the project's problems are widely appreciated by others.

But the top-management of the firm will not be ignorant of this logic, although they cannot identify which manager or which project is opportunistic. They are forced to distrust and therefore discount all claims about future profits even more sharply, further increasing institutional myopia.

Now consider the dilemma faced by a midlevel manager who actually has a valuable entrepreneurial idea. The organization, rationally responding to the problems of mobility and opportunism, discounts the longer-term aspects of the proposal or presses for greater collective support by higher-level managers. Seeing that either the project will be rejected or future credits for success will be shared with powerful superiors, the entrepreneurial manager has incentives to leave the firm and pursue the project independently, if possible. By leaving the firm and substituting an ownership interest for an employment relationship, the entrepreneur increases his ability to bond his word by placing his own wealth at risk[9] and providing contractual and organizational arrangements that more tightly link future returns with his wealth or reputation.

To close the analysis, it must be noted that the entrepreneur's ability to exit the firm and form his own venture is yet another avenue that *increases* the myopia within the firm. Given this alternative, the senior management must consider that any proposal they receive is one that would not be acceptable to the external venture capital market!

The above theory accomplishes two things. It provides an explanation for institutional myopia wherein all actors are rational, and it explains exits and spin-offs in terms of incentive failure rather than as intellectual theft. That is, phenomena like Silicon Valley, where a multitude of firms are formed by employees who quit and take ideas to venture capitalists, can be understood as solutions to the problems of incentives within firms.

Conclusions

Entrepreneurial activity will be encouraged where appropriability is low and isolating mechanisms are high. These areas may not necessarily be those where the social returns to innovation are highest, but they are those where private returns to innovation exist. The connection between entrepreneurial activity and uncertainty cannot be signed in general, but there are reasons to believe that it may be positive in many cases.

Entrepreneurship within organizations is facilitated by the ability of large firms to muster resources and administer large projects; it is inhibited by bureaucratic inertia and by the incentive problems rising out of informational asymmetries.

With regard to the organizational locus of entrepreneurship, the analysis points up the salience of the project's futurity. As more of the expected returns to investment occur in the distant future, the potential entrepreneur's ability to attract investors diminishes. Coupling this notion with the problem of appropriability by cospecialized assets, the diagram shown in Figure 7–1 may be constructed. Where the entrepreneurial venture involves significant cospecialized assets, the expectation is that it will be undertaken by a firm possessing those assets. However, as the

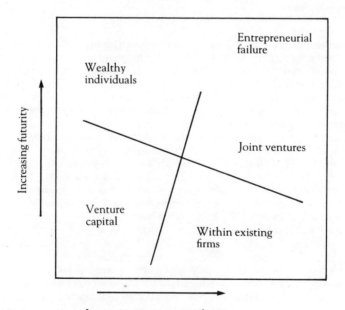

Figure 7–1. Variations in Entrepreneurial Locus with Futurity and Degree of Asset Cospecialization

project's futurity increases, it becomes more difficult to assess the project within the firm. Without some external bonding mechanism, entrepreneurial failure may ensue. One possible solution is a joint venture, with both the cospecialized asset owner and the entrepreneurial visionary investing in a new separate corporation.

Where cospecialized assets are not a problem, we expect to see new ventures formed by individuals, some of whom may be exiting from firms that are unable to provide the appropriate incentive arrangements. Still, when the futurity of these projects grows too large, the problem of obtaining resources can become insurmountable. In this final region, entrepreneurship will be the province of those who have the ideas and are already wealthy enough to indulge them.

NOTES

1. Griliches's (1984) book provides a fine compendium of recent work in the field.
2. Taken from Rumelt (1982).
3. Historically, the term *rent* applies to continuing nondiminishing payments. Above-normal returns that diminish over time are frequently labeled *quasi-rents*. However, modern theory is less concerned with long-term equilibria and more concerned with the *ex ante* equilibria of expectations. In this context, in which values are present values rather than annuities, we use the simple term *rent* to cover both quasi-rents and persistent rents.
4. It is sometimes useful to distinguish between appropriability with respect to use and general appropriability. If, for example, land earns rents in beet farming and would earn equivalent rents in bean farming, taxes on beet production cannot appropriate the rents because the farmer can simply switch to raising beans. Thus, appropriability with respect to use depends on the specificity of the resource—its relative value in its best use compared with that in its next best use. By contrast, a general tax on land income could appropriate the land's rent regardless of its use.
5. More precisely, if the entrepreneur faces uncertainty represented by the random variable X with distribution function F and has a convex payoff function $V(x)$, where x is a realization of X, then $EV(x)$ increases with mean-preserving increases in the riskiness of F (in the sense of second-order stochastic dominance). A more familiar application of this result is in option theory, where it is well known that increases in the variance of the underlying security increase the values of existing options.
6. See Kamien and Schwartz (1982) and Reinganium (1983).
7. I am assuming that once a manager takes a position in another organization there is a halt to the process of updating his reputation based on results occurring within his previous employer's organization. Mobility is the simplest way in which this type of "forgetfulness" can be invoked. Other mechanisms commonly recognized in practice are promotion, reorganization, moving to a different division, changes in accounting, the bundling or unbundling of projects, and changes in senior management. In each case, the connection between the project that was once "sold" by the manager and updates to the manager's reputation is weakened.
8. I am implicitly assuming that the market for managerial labor is a lemons market, in the sense of Ackerlof (1970). That is, the market exists because there are nonoppor-

tunistic reasons for managerial mobility so that "lemons" can, if their quantity is sufficiently low, mix in with the good-quality managers.
9. Were the original employer not large and diversified, he could have accomplished this within the firm simply by buying stock.

REFERENCES

Akerlof, G.A. 1970. "The Market for 'Lemons': Quality Uncertainty and the Market Mechanism." *Quarterly Journal of Economics* 84: 488–500.

Arrow, K. 1962. "Economic Welfare and the Allocation of Resources for Inventions." In *The Rate and Direction of Inventive Activity*, edited by R.R. Nelson. Princeton, N.J.: Princeton University Press.

Bower, J.L. 1967. *Managing the Resource Allocation Process.* Cambridge, Mass.: Harvard University Press.

Caves, R.E., and M.E. Porter. 1977. "From Entry Barriers to Mobility Barriers: Conjectural Decisions and Contrived Deterrence to New Competition." *Quarterly Journal of Economics* 91 (May): 241–61.

Comanor, W.S. 1967. "Market Structure, Product Differentiation, and Industrial Research." *Quarterly Journal of Economics* 81: 639–57.

Conner, K., and R.P. Rumelt. 1986. "Software Piracy." UCLA working paper.

Crozier, M. 1964. *The Bureaucratic Phenomena.* Chicago: University of Chicago Press.

Demsetz, H. 1969. "Information and Efficiency: Another Viewpoint." *Journal of Law and Economics* 12 (April): 1–22.

Downs, A. 1967. *Inside Bureaucracy.* Boston: Little, Brown.

Downs, D.H., and R. Heinkel. 1982. "Signaling and Valuation of Unseasoned New Issues." *Journal of Finance* 37: 1–10.

Drucker, P.F. 1985. *Innovation and Entrepreneurship.* New York: Harper & Row.

Galbraith, J.K. 1952. *American Capitalism: The Concept of Countervailing Power.* Boston: Houghton Mifflin.

Grayson, C.J., Jr. 1960. *Decisions under Uncertainty: Drilling Decisions by Oil and Gas Operators.* Boston: Harvard Business School.

Griliches, Zvi, ed. 1984. *R&D, Patents, and Productivity.* Chicago: University of Chicago Press.

Hitt, M.A., and R.D. Ireland. 1985. "Corporate Distinctive Competence, Strategy, Industry and Performance." *Strategic Management Journal* 6 (July/September): 273–93.

Jacobson, G., and J. Hillkirk. 1986. *Xerox: American Samurai.* New York: Macmillan.

Kamien, M.I., and N.L. Schwartz. 1982. *Market Structure and Innovation.* Cambridge: Cambridge University Press.

Klein, B.K., and K.B. Leffler. 1981. "The Role of Market Forces in Assuring Contractual Performance." *Journal of Political Economy* 89 (August): 615–41.

Leland, H., and D. Pyle. 1977. "Informational Asymmetries, Financial Structure, and Financial Intermediaries." *Journal of Finance* 32: 371–87.

Lippman, S.A., and R.P. Rumelt. 1982. "Uncertain Imitability: An Analysis of Interfirm Differences in Efficiency Under Competition." *Bell Journal of Economics* 13 (Autumn): 418–38.

Mansfield, E. 1968. *Industrial Research and Technological Innovation.* New York: Norton.

———. 1971. *Research and Innovation in the Modern Corporation.* New York: Norton.

Mintzberg, H., D. Raisinghani, and A. Theoret. 1976. "The Structure of 'Unstructured' Decision Processes." *Administrative Science Quarterly* 21 (June): 246–75.

Nelson, R.R., and S.G. Winter. 1982. *An Evolutionary Theory of Economic Change.* Cambridge, Mass.: Belknap Press of Harvard University Press.

Phillips, A. 1971. *Technology and Market Structure: A Study of the Aircraft Industry.* Lexington, Mass.: Lexington Books.

Reinganium, J.F. 1983. "Uncertain Innovation and the Persistence of Monopoly." *American Economic Review* 73 (September): 741–48.

Ricardo, D. 1971. *Principles of Political Economy.* Edited by R.M. Hartwell. London: Penguin.

Rumelt, R.P. 1982. "How Important Is Industry in Explaining Firm Profitability." UCLA working paper.

———. 1984. "Towards a Strategic Theory of the Firm." In *Competitive Strategic Management,* edited by R.B. Lamb, pp. 566–70. Englewood Cliffs, N.J.: Prentice-Hall.

Scherer, F.M. 1965. "Firm Size, Market Structure, Opportunity, and the Output of Patented Inventions." *American Economic Review* 55: 1097–125.

———. 1967. "Research and Development Resource Allocation under Rivalry." *Quarterly Journal of Economics* 81: 524–31.

Schon, D.A. 1967. *Technology and Change.* New York: Delta.

Schumpeter, J.A. 1950. *Capitalism, Socialism, and Democracy,* 3d ed. New York: Harper & Row.

Teece, D.J. Forthcoming. "Profiting from Technological Innovation." *Research Policy: Implications for Integration, Collaboration, Licensing, and Public Policy.*

Wernerfelt, B. 1984. "A Resource-Based View of the Firm." *Strategic Management Journal* 5 (April-June): 171–80.

Williamson, O.E. 1965. "Innovation and Market Structure." *Journal of Political Economy* 73: 67–73.

8 KNOWLEDGE AND COMPETENCE AS STRATEGIC ASSETS

Sidney G. Winter

An *asset*, my dictionary says, may be defined as "a useful thing or quality." Among commentators on corporate strategy, it is widely accepted that knowledge and competence are useful things for a company to have. At times, particular approaches to the acquisition and profitable exploitation of productive knowledge—such as the experience curve—have been the central focus of strategic discussion. At other times, explicit attention to the place of knowledge considerations in the strategic picture has waned, perhaps to the point where such issues have "dropped through the cracks" of strategic analysis (see Peters 1984: 115). But they certainly cannot drop very far below the analytical surface because any discussion of innovation and indeed the activity of strategic analysis itself implicitly concedes their importance.

The dictionary offers an alternative definition of an asset as "a single item of property." In some cases, this second meaning may be applicable, along with the first, to knowledge held by a business firm. A basic patent, for example, may certainly be a useful thing for a company to have, and at the same time it may represent a discrete bundle of legally defined and enforceable property rights; such an item of property can be conveyed

I am indebted to seminar participants at the University of California, Berkeley, and the University of Pennsylvania for helpful comments and challenging questions. David Teece gets special thanks both for his suggestions and for his patient encouragement of the entire effort (without thereby incurring any responsibility for the result). Financial support from the National Science Foundation, Division of Policy Research and Analysis, and from the Sloan Foundation is gratefully acknowledged, as is the research assistance of Allen Presseller.

from one owner to another just as a stock certificate or a deed can. In general, however, it is decidedly problematic whether the realities denoted by such terms as *knowledge, competence, skills, know-how,* or *capability* are the sorts of things that can be adequately discussed as items of property. The word *item* is suggestive of a discreteness, of a potential for severance from the prevailing context, that is frequently not characteristic of the skills of individuals and organizations. The term *intellectual property* is established in legal parlance, but there are nevertheless often profound ambiguities in both principle and practice regarding the scope and locus of the rights associated with the possession of knowledge.

Thus, of the two definitions of *assets,* one is plainly applicable to the knowledge and competence of a business firm, while the other is of uncertain applicability. This situation does not pose a problem for lexicographers, but it does pose a problem for analysts of strategy. The reason is that the disciplines of economics, accounting, and finance have developed and defined the asset concept in ways that are largely specializations of the second dictionary definition (which, of course, is itself a specialization of the first definition). Where the second definition does not apply, the tool kits of those important disciplines contain little that affords a useful analytical grip on strategic issues. Systematic analysis is crippled, and many important issues are addressed only with the general purpose tools of aphorism and anecdote.

This chapter attempts to bridge the gap between the two meanings of *asset* as they relate to the knowledge and competence of a business firm. The first section introduces a state description approach to strategy that borrows elements from optimal control theory and from evolutionary economics. The next two sections develop this approach, with particular emphasis on its relationship to the valuation of the firm's productive system, link the state description approach to the diverse set of organizational phenomena denoted by *knowledge, competence,* and kindred terms, and explore some distinctions among these phenomena that are of major importance for strategy. The next section draws on the Yale survey of corporate R&D managers to suggest the extent to which the key mechanisms affecting the creation and diffusion of productive knowledge differ from one branch of manufacturing industry to another. The final section briefly reviews the major themes of the chapter.

Strategic State Description

An organizational strategy, I propose, is a summary account of the principal characteristics and relationships of the organization and its environment—an account developed for the purpose of informing decisions affecting the organization's success and survival. This formulation emphasizes the normative intent of strategic analysis and rejects the notion that there are strategies that have "evolved implicitly" (Porter 1980: xiii) or that strategy is a "non-rational concept" (Greiner 1983: 13). There may of course be strategies that are clearly formulated but then rejected or ignored, that are not written down but are nevertheless successfully pursued, or that lead to abject failure rather than to success. These realistic possibilities are consistent with a view of strategic analysis as a form of "intendedly rational" (Simon 1957: xxiv) behavior directed toward pragmatically useful understanding of the situation of the organization as a whole. By contrast, mere habits of thought or action, managerial or otherwise, are not strategies. Such habits may as easily be parts of an (unintended) problem as parts of an (intended) solution.

The propensity to perceive habits as strategies may well derive from the valid observation that much of the behavior of an organization is quasi-automatic and neither requires the attention of top management on a day-to-day basis nor, when it receives such attention, responds to it in a straightforward and constructive way. This phenomenon is most frequently noted in connection with organizational resistance to change—that is, as a problem facing change agents. But it is also cited as a factor on the bright side, contributing to excellence in organizational performance or as indicative of the positive results achievable when lower levels of the organization are successfully imbued with appropriate operational versions of organizational goals. Thus, as Peters (1984: 111) says:

> Distinctive organizational performance, for good or ill, is almost entirely a function of deeply engrained repertoires. The organization, within its marketplace, *is* the way it *acts* from moment to moment—not the way it thinks it *might* act or *ought* to act.

Peters's statement is quite consistent with the viewpoint on organizations that is basic to the evolutionary economic theory developed by Richard Nelson and myself (Nelson and Winter 1982: 134–36.) The same view is a fundamental constituent of the approach to strategy set forth in this chapter. Nevertheless, it should be clear that this view is potentially quite subversive of the whole undertaking of normative strategic analysis. The more deeply ingrained the organizational repertoires, the less clear it is what important decisions remain for an analyst to advise on or an executive to execute. When apparent choice situations of apparent strategic significance confront the organization, perhaps outcomes are fully determined by some combination of habit and impulse.[1]

This is exactly the way the organizational world is envisaged, for purposes of *descriptive* theorizing, in evolutionary economics. In that theoretical world, strategic analysis in the sense defined here has no place, although of course there is abundant scope for *ex post facto* discussion of which habits and impulses proved successful. As a response to a need for guidance in the real world, this fatalistic perspective has obvious and severe limitations.

A key step in strategic thinking is the identification of the attributes of the organization that are considered subject to directed change and the implicit or explicit acknowledgment that some attributes do not fall in that category. As the Alcoholics Anonymous serenity prayer puts it, "God grant me the serenity to accept the things I cannot change, courage to change the things I can, and wisdom to know the difference." Substitute for *courage* the words *managerial attention and related resources supporting strategic decisionmaking* and you have here the beginnings of a paradigm for strategic analysis, its role being to help with the wisdom part. Of course, the sort of wisdom contributed by an economist will include the observations that change *per se* is presumably not the goal, that change will often be a matter of degree, and that the trick is to allocate the available change capacity in the right way.

State Description as a Generalized Asset Portfolio

The concept of the state of a dynamic system has a long history in control theory and related subjects (Bellman 1957; Pontryagin *et al.* 1962; and a vast literature of applications). The distinction between the *state variables* of a system and the *control variables* is roughly the distinction between aspects of the system that are not subject to choice over a short time span and aspects that are. The values chosen for control variables, however, do affect the evolution of the state variables over larger time spans. As far as the internal logic of a control theory model is concerned, the list of state variables constitutes a way of describing the system that is sufficiently precise and comprehensive so that the motion of the system through time is determined, given the settings of the control variables and the state of the external environment at each point of time.

In general, a variety of alternative state descriptions provide formally equivalent approaches to a given problem. Also, the conceptual distinctions among state variables, control variables, and the environment can become blurred in the sense that particular considerations may be treated under different headings in formulations of two very closely related problems. There is, for example, little substantial difference between a given feature of the environment and a system state variable that is alterable only over a very narrow range.

In evolutionary economics, the notion of state description is extended to cover behavioral patterns that most economists or management scientists would instinctively place in the control variable category. Behavior is conceived as governed by *routines* (or alternatively, by "deeply ingrained repertoires") rather than deliberate choice. The object of a theoretical exercise is not to discover what is optimal for a firm, but to understand the major forces that shape the evolution of an industry (Nelson and Winter 1982: 18–19):

> The core concern of evolutionary theory is with the dynamic process by which behavior patterns and market outcomes are jointly determined over time. The typical logic of these evolutionary processes is as follows: At each point of time, the current operating characteristics of firms, and the magnitudes of their capital stocks and other state variables, determine input and output levels. Together with market supply and demand conditions that are exogenous to the firms in question, these firm decisions determine market prices of inputs and outputs. The profitability of each individual firm is thus determined. Profitability operates, through firm investment rules, as one major determinant of rates of expansion and contraction of individual firms. With firm sizes thus altered, the same operating characteristics would yield different input and output levels, hence different prices and profitability signals, and so on. By this selection process, clearly, aggregate input and output and price levels for the industry would undergo dynamic change even if individual firm operating characteristics were constant. But operating characteristics, too, are subject to change, through the workings of the search rules of firms. Search and selection are simultaneous, interacting aspects of the evolutionary process: the same prices that provide selection feedback also influence the directions of search. Through the joint action of search and selection, the firms evolve over time, with the condition of the industry in each period bearing the seeds of its condition in the following period.

It is clear that *among* the things that are candidate variables inclusion for a state description of a business firm are the amounts of the firm's tangible and financial assets, the sorts of things that are reflected on the asset side of a balance sheet. It is equally clear that the conception of a firm state description in evolutionary theory goes well beyond the list of things conventionally recognized as assets. Theoretical studies employing control theory techniques, by economists and management scientists, have established characteristics of optimal behavior in problems in which state variables correspond closely to things recognizable as assets—for example, inventory or capacity levels—but also things that are not so recognized, at least in financial accounting—for example, stocks of customers, employees, and advertising or R&D capital. There is, therefore, a relationship but also a conceptual gap between the concepts of a *state description* and a collection (or *portfolio*) of assets. There is likewise a relationship but also a gap between evolutionary theory's notion of a state description for a *business firm*, a description that is comprehensive in principle, however limited or stylized it may be in a particular analytical application, and the descriptions that derive from the conventions of asset accounting or from the focused objectives that necessarily govern the construction of a control theory model, whether for theoretical or practical use. In particular, the state description concept in evolutionary theory, and the concept of a routine more specifically, direct attention to the problem of reflecting the knowledge and competence of a firm in a state description—but offer only minimal guidance as to how this might be done.

Organizational Goals

The bridges that are to be constructed across the gaps just referred to must be anchored, at least for the time being, in a strong commitment regarding the goals of the organizations whose strategic problems are to be analyzed. The discussion will relate only to organizations for which present value maximization, or expected present value maximization, adequately characterizes the organizational goal as perceived by the actor or actors for whose guidance the analysis is conducted. It is to be hoped that some of the illumination will extend well beyond the range of the assumptions adopted here, but how far that may be the case is an issue that must be left open.[2]

The assumption of a present-value goal for the organization places this analysis in a simple and orthodox tradition in economic theorizing. This tradition of viewing the firm as a unitary actor with well-defined preferences has long been challenged by organization theorists and social scientists outside of economics, and by a few economists of heretical bent (such as Cyert and March 1963). Increasingly, this tradition has been abandoned by numerous theoretical economists of diverse points of view, and the assumption made here might well be regarded as a throwback.[3] There are indeed some key issues in the strategic management of knowledge assets that relate to whether the firm can hold together in the face of conflict among the diverse interests of the participants. Although these issues are touched on below, for the most part they remain on the agenda for future work. The assumption that the present value of the concern is maximized is maintained for the time being in the spirit of dividing the difficulties.

The major restrictions on the scope of this discussion having been duly noted, it is now time to emphasize its generality. The concept of a system state is highly flexible, yet within the confines established by a present value criterion it can easily be linked to the conventional (second definition) concept of an asset. A complete and accurate state description for a business firm is plainly an unattainable goal outside the confines of a theoretical model. Yet the idea of seeking a normatively useful state description is realistic and familiar. When the normative purpose in view involves the direction of the entire organization, an attempt at strategic state description may be helpful.

Theories of strategy, accounting principles, and many other aspects of business practice can be understood as providing conceptual structures for state descriptions that are practical and often quantitative but clearly partial. The value of these schemes, at least regarding the strategic guidance they provide, is often limited by the weakness of their connections to economically relevant conceptions of assets and returns on assets. Relatedly, and perhaps more significantly, they are but weakly connected to the most basic of all paradigms for making money—"buy low, sell high."[4] The strategic state description paradigm developed here does not suffer from these limitations, but, as emphasized below, it cannot escape the fact that any implementable state description scheme is necessarily partial.

State Description and Valuation

Full Imputation

The mathematics of optimal control theory reflects a long-familiar heuristic principle in economic thinking, the principle of *full imputation*. This principle states that a proper economic valuation of a collection of resources is one that precisely accounts for the returns the resources make possible.[5] For present purposes, the simplest relevant application of this principle is "an asset (def. 2) should be valued at time at T at the present value of the net returns it will yield from T onward." A more exact formulation, appropriate for present purposes, is that the owner(s) of an asset should value it at the present value of the net future returns it generates under present ownership, where the interest rate(s) employed in the discounting reflect the lending opportunities open to the owners. This is an owners' reservation price valuation; if more than this is offered for the asset, the owners should take it. In the optimal control theory context with a present value criterion, a more complex version of this same proposition attributes the maximized present value attainable from the system to its initial state together with features of the environment and the laws of change. Of course, the policy choices made affect the present value achieved—but since *optimal* control theory points the way to *optimal* choices of these policy variables, once these choices are made the policy followed is not explicitly a determinant of value.

The adoption of this valuation principle carries the direct implication that the notion that an excess return or (economic) profit can be earned by holding an asset is illusory. Properly valued assets yield only normal returns, where *properly valued* refers to the owner's reservation price defined above. If there is a gain or loss, the full imputation principle declares it to be a *capital* gain or loss associated with having acquired the asset at a price below or above its true value—that is, it is in the nature of a success or failure in speculation.

What sort of speculation is involved, and what are the sources of success in this activity? One clear possibility is blind luck in the making of decisions to buy or sell. Perhaps success also can be explained by superior knowledge, competence, insight, skill, or information.[6] But the guidance provided by the full imputation principle suggests that the words *can be explained by* might reasonably be replaced by *should be imputed to*— a conceptual maneuver that makes the full imputation fuller than it was before, restores blind chance as the sole source of net returns, and leaves us with a conception of the assets involved in the situation that is broader and more remote from financial accounting conventions than it was before. The subtleties of this imputation dialectic—full imputation for one process discloses unaccounted returns in a casually antecedent process, which then calls for a fuller imputation—must be confronted if strategic analysis is to have solid foundations in economic reasoning.

Whether confronted or not, they are key issues when the strategic options include acceptance or rejection of a bid to purchase the company, and in the wider range of cases involving transactions in functioning business units, large or small. Rational action in such situations demands attention to the question of what future earning power actually "comes with" the entity whose ownership is transferred. That question cannot be answered without inquiring deeply into the sources of earning power— that is, without confronting the imputation problem.

The subtleties are particularly fundamental to understanding the strategic role of knowledge and competence. For, as is discussed further in the next section, policies affecting the growth or decline of knowledge and competence assets can have major effects on earning power over time but may do so without posing the question "What is this worth?" with the clarity with which it is posed in a major transaction.

State Descriptions, Optimization, and Heuristic Frames

Strategic analysis would present no challenge if only two conditions were satisfied; (1) if it were easy to identify the real problem faced by the organization (that is, to correctly identify and assess the state variables, control variables, constraints, and laws of change affecting the organizational system) and (2) if the problem thus identified were easily solved. Once these easy steps were taken, any apparent superiority or inferiority in actual organizational performance over time would reflect the play of pure chance. In fact, because of the bounded rationality of individuals and organizations, neither of these conditions is remotely satisfied in the real world. It is therefore inevitable that real strategic analysis involve highly simplified and perhaps fragmented conceptualization of what the strategic problem is and that the solution to the identified problem involve a continuing process of situational analysis, decisionmaking, action taking, and evaluation of the results. Difficulties in implementation appear, which is to say that some of the things conceived as control variables do not have the anticipated effects on things conceived as state variables or that presumed control variables themselves turn out to be not so controllable after all. Surprises occur as environmental situations arise that were not conceived as possible or were regarded as of negligibly low probability. Failures to comprehend fully the internal logic of the strategic problem may become manifest in coordination failures and intraorganizational conflict.

Since strategic analysis is necessarily imperfect in the real world, there is always room for improvement. In general, there is room for improvement both at the stage of problem definition (since the real problem, nicely formulated, is not handed to the analyst on a platter) and at the stage of problem solution (since it is not necessarily a trivial matter to derive the policy implications of a statement of the strategic problem, however clearly formulated). Because bounded rationality limits achievement at both stages, the relations between the two are more subtle than the simple define problem/solve problem scheme might suggest. Stage one must be conducted with a view both to capturing the key features of the strategic situation and with a view to the available capabilities for deriving specific conclusions in stage two. As a result, there is a tension or tradeoff between flexibility and scope on the one hand versus problem-solving power on the other.

I will use the term *heuristic frame* to refer to a collection of possible approaches to a particular strategic problem whose members are related by the fact that they all rely on the same conception of the state variables and controls that are considered central to the problem. A heuristic frame corresponds to a degree of problem definition that occupies an intermediate position on the contiuum between a long and indiscriminate list of things that might matter at one end and a fully formulated control-theoretic model of the problem at the other. Within a heuristic frame, there is room for a wide range of more specific formulations of the problem—but there is also enough structure provided by the frame itself to guide and focus discussion. On the other hand, a rich variety of different heuristic frames may represent plausible approaches to a given problem. Commitment to a particular frame is thus a highly consequential step in strategic analysis and one that deserves careful consideration.

Most of the approaches to strategic problems that are to be found in the literature do not involve explicit reference to state descriptions and heuristic frames or emphasize the possibility of translating the analysis into the language, if not the formalism, of control theory. (Therein lies, of course, the claim to novelty of the strategic state description approach presented here.) Many strategic perspectives can, however, be recast in this form without too much sacrifice of content and often with the benefit of revealing gaps, limitations, or vagueness in the particular perspective. The danger of neglecting alternative heuristic frames may also be highlighted.

Consider, for example, the classic BCG doctrine based on the experience curve. At the level of the individual line of business, this doctrine identifies unit cost and cumulative output to date as the key state variables and current output as the control. The connection from output to unit cost is mediated by cumulative output and the experience curve; an obvious identity relates current output and cumulative output to the new value of cumulative output. Market share can seemingly play alternative roles in the scheme, being sometimes a surrogate for output as the control variable (especially in situations in which multiple outputs are involved), sometimes a surrogate for cumulative output as a determinant of cost (where shares have been constant over an extended period), sometimes a control variable causally antecedent to current output (where the step between producing more output and getting it sold is itself strategically problematic), and sometimes a state variable subsequent to unit cost (low cost makes high share attainable, and high share makes high profits attainable). At the corporate level, the state description is the list of lines of business, with each line characterized by a market growth rate and a market share. Allocations of investment funds to the various lines of business are among the control variables. An allocation to a line of business relieves the cash constraint on the control variable for that line, making possible an increase of market through a capacity increase, advertising campaign, price cut, or whatever. Acquisitions and divestitures are also corporate level controls, making possible direct changes in the corporate-level state variables.

The general character of the normative guidance that is loosely derived from this scheme presumably needs no review. One notable feature of most accounts of the BCG approach in the literature is the absence of discussion of the costs at which changes in state variables are affected, at either the line of business or corporate level; needless to say, there is also no discussion of balancing costs and benefits at the margin. A second feature is the sparsity of the description of the environment, which treats market share as the only significant characteristic of rivals and also does not address features of the market itself that, along with growth, might affect profitability—such as the price elasticity of demand. Spelling out the heuristic frame of the BCG analysis, and noting the character of the questions that would need to be answered to complete a control-theoretic formulation thus leads rather quickly to the identification of what seem to be important gaps; it is proposed that the same critical approach might be helpful more generally.

Applying the full imputation principle within the BCG heuristic frame tells us what the value of a company depends on. It depends on the unit cost and cumulative output levels in all the individual lines of business, considered in conjunction with the environmental facts of market sizes, market growth rates, and sizes of leading rivals—plus, of course, the net financial assets of the company. If the heuristic frame were fully and correctly expanded into an optimal control problem, and that problem were in turn correctly solved, then (for the purposes of strategic analysis) these considerations and net financial assets would be the *only* determinants of the company's value; nothing would be left of the strategic problem but to carry out the optimal policy. Actually, since problem formulations and solutions are the imperfect products of bounded rationality, there will almost certainly be room for influencing the value of the company through a different implementation of the strategic approach defined by a particular heuristic frame.

More central to the purposes of the present paper is the observation that a quite different heuristic frame might be adopted. An alternative frame provides a different list of things that influence profitability in the long run and of how these relate to things that are controllable in the short run. A different approach to valuation, and perhaps a very different result, is implied. Whether the new strategic valuation is lower or higher is not indicative of the merit of the change of frame; what matters is whether the guidance obtained from the new frame serves the company better or worse in obtaining actual returns than the guidance obtained from the old. In particular, a change of heuristic frame may be a response to recognition that the old frame embodies an overoptimistic view of the strategic situation and the present value it implies is unrealizable.

Describing Knowledge States

Simple descriptors of knowledge states, often involving a single variable, have played an important role both in the theory of economic growth and in empirical research on R&D, the determinants of profitability, and related topics.[7] Although considerable insight has been derived from these studies, both the theory and the evidence are generally at too aggregative a level to be more than the suggestive for the purposes of strategic analysis. The domain of strategic choice certainly includes, for example, the choice of a level of R&D expenditure, but such a choice ordinarily interacts with the details of project selection.

It is therefore necessary to confront the difficulties that arise from the complexity and diversity of the phenomena denoted by such terms as knowledge, competence, skill, and so forth. When we use such terms, we hardly ever know precisely what we are talking about (except when we are expert in the area under discussion), and there is sometimes a nagging concern that we are too far from the complex details to be making sense at all. The purpose of this discussion is to alleviate this situation in some degree—to introduce some distinctions that clarify the conceptual issues surrounding knowledge and competence as strategic assets.

Taxonomic Dimensions

Suppose that we have under discussion something that we tentatively think of as a knowledge or competence asset. Figure 8–1 below lays out some dimensions along which we could try to place this asset and thus come to a clearer understanding of what the thing is and what its strategic significance might be. In general, a position near the left end of any of the continua identified in the figure is an indicator that the knowledge may be difficult to transfer (thus calling into question its status as an asset in the second sense), whereas a position near the right end is indicative of ease of transfer. This interpretation is elaborated below.

The first of the continua listed in Figure 8–1 ranges from highly tacit to fully articulable knowledge. Individual skills are often highly tacit in the sense that *"the aim of a skillful performance is achieved by the observance*

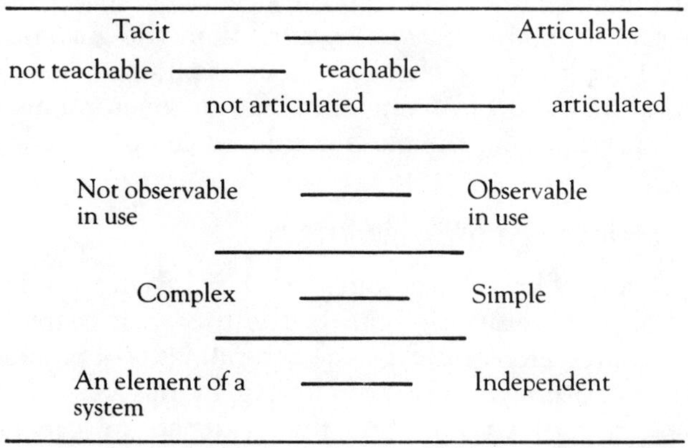

Tacit	————	Articulable
not teachable	———— teachable	
	not articulated ————	articulated
Not observable in use	————	Observable in use
Complex	————	Simple
An element of a system	————	Independent

Figure 8–1.   Taxonomic Dimensions of Knowledge Assets

of a set of rules which are not known as such to the person following them" (Polanyi 1962: 49, emphasis in original). "Not known as such" here means that the person could not provide a useful explanation of the rules. Fully articulable knowledge, on the other hand, can be communicated from its possessor to another person in symbolic form, and the recipient of the communication becomes as much "in the know" as the originator.

The reality of the phenomenon of tacit knowing at the level of individual skills is obvious from introspective evidence; its sources and significance were explored in depth in Nelson and Winter (1982). An article in the *New York Times* (Blakeslee 1985) cited recent scientific evidence that different brain structures are involved in memory for the "procedural" knowledge underlying skills as opposed to memory for the "declarative knowledge" of facts, and provided a striking example of the distinction. A brain-damaged man retained his ability to play a good game of golf (something that he obviously could not transfer to another person by mere communication), but could not recall where the ball had just landed or keep track of his score (fully articulable knowledge that his damaged memory could not retain).

Knowledge possessed by an organization may be tacit knowledge in the sense, first, that the possession arises from the association with the organization of an individual for whom the knowledge in question is tacit. Related articulable knowledge may be possessed by other members of the organization, to the effect that "We have someone who knows about (or can do) that." Second, the fact that the myriads of relationships that enable the organization to function in a coordinated way are reasonably understood by (at most) the participants in the relationship and a few others means that the organization is certainly accomplishing its aims by following rules that are not known as such to most participants in the organization. Third, in a metaphorical sense an organization's knowledge is tacit to the extent that its top decisionmakers are uninformed regarding the details of what happens when their decisions are implemented. The decisionmakers are the symbol-processing brains of the organization; the symbols they deal in may suggest very little of what the nerves, bones, and muscles actually do—even though, in the reality to which the metaphor relates, the nerves, bones, and muscles may be quite capable of describing it to each other.

Tacit skills may be teachable even though not articulable. Successful teaching presupposes the willingness of the pupil to engage in a series of trial performances of the skill and to attend to the teacher's critique of the errors made in these trials. Teachers may also provide model performances of the skill, which provide the pupil with an opportunity for imitative learning. Instruction of this sort may accomplish a radical reduction in the time and effort required for skill acquisition, relative to what would be required by the pupil proceeding on trial and error alone—but the situation nevertheless is vastly different from one in which knowledge is fully conveyed by communication alone.

A second subdimension identified in Figure 8–1 is the distinction between articulable knowledge that *is* articulated and articulable knowledge that is not. The latter situation may be illustrated by the case of a complex computer program that has gone through a number of major revisions since its documentation was last brought up to date. Simple answers to questions about the program's functioning could be articulated in principle but may not be articulable in fact if the information is no longer in someone's memory. Similar situations seem to arise frequently in manufacturing, where the actual process or product design being followed deviates systematically from the symbolically recorded design or plan. Personnel turnover or simple lapse of time may erase the organization's memory of the actual process or design if the production routine is not regularly exercised and thus remembered by doing. A related phenomenon is that a deviation from the nominal standard may remain unarticulated because it includes features that are to the advantage of an individual organization member but perhaps not to the organization as a whole.

In Figure 8–1, the fact that the not articulated position is placed to the left of the teachable position is intended to suggest the point that the failure to articulate what is articulable may be a more severe handicap for the transfer of knowledge than tacitness itself.

Observability in use, the second of the major continua in Figure 8–1, involves the extent of disclosure of underlying knowledge that is necessitated by use of the knowledge. The design of a product is a secret that is hard to keep if the product is made available for purchase (and inspection) by all comers in an open market. In general, the question at issue involves the opportunities that use makes available to someone who wishes to discover the underlying knowledge. The resources that such an individual has to apply to the task, relative to the costs of observation, should be taken into account in operationalizing this conceptual dimension. Also, the question of whether the observation is taking place with or without the cooperation of the organization or individual observed is a key contextual feature of the discovery task.

The complexity/simplicity dimension has to do with the amount of information required to characterize the item of knowledge in question. Here, as elsewhere in information theory, the notion of an amount of information must be interpreted in terms of the alternative possibilities from which a particular case must be distinguished. For example, the item in question might superficially have to do with the design of an

automobile. But perhaps everything about the design is familiar throughout the relevant context except the ceramic material used in the spark
plugs, which itself is distinguished from possible alternatives by its name.
In that context, the apparently complex is actually simple.

Similar issues arise in connection with the final dimension. A single
module in a microcomputer qualifies intuitively as an element of a system.
A pocket calculator is at the opposite end of the spectrum; it is useful
standing alone. In a context where all other elements of the microcomputer system are readily available, however, the individual module might
be said to be useful by itself.

Strategic Significance

As suggested above, the left-hand ends of the continua in Figure 8–1
are unfavorable to knowledge transfer: Transfer of tacit knowledge, if
possible at all, requires teaching; an element of a system may not be
helpful if transferred without the rest of the system, and so forth. Ease
of transfer is itself a decidedly ambiguous variable. From the strategic
point of view, it is crucial here to distinguish *voluntary* from *involuntary*
transfers of knowledge. Among the most important peculiarities of knowledge and competence as assets is that secure control of such assets is
often very difficult to maintain. No one can walk out the gate of a steel
plant or a refinery taking the economic value of the physical installation
with him in his pocket, leaving a hollow shell behind. The same is not
true of an R&D lab, since the pocket may contain an articulated statement
of a simple item of knowledge whose value is substantially independent
of the value of other knowledge that remains behind in the lab. And
even though what is in the pocket may be only a copy of something that
remains in the lab, it may suffice to make the original a hollow shell
without economic value.

A recent *Fortune* article (Flax 1984) provides a handy list of twenty-one
ways that companies snoop on their rivals, ranging from such familiar
methods as hiring away rivals' employees and reverse engineering their
products to more esoteric or exotic techniques such as getting customers
to put out phony bid requests and buying competitors' garbage for analysis.
Because it focuses on "snooping" activity, the list omits one major route
by which knowledge may escape from control, the "fissioning" of the
company as new, entrepreneurial enterprises are founded by its former
employees.[8] For the same reason, it also omits the important category
of voluntary disclosure through patent applications, advertising, and contract bidding.

The key strategic questions are (1) What sorts of knowledge and competence assets are worth developing and (2) how is value to be derived

from those assets? As will be emphasized below, intrinsic differences among knowledge bases and other circumstances of different areas of technology and organization are important determinants of where newly developed assets tend to fall along the taxonomic dimensions identified above. To the extent this is true, the implications for strategic choice relate to which areas to be in. These implications in turn must be assessed in light of the previous section's emphasis on the initial state of the system as a fundamental determinant of value: If the company in question is in the toy business, it is not helpful to observe that the prospects for protecting knowledge assets are better in the chemical business. Less dramatic transformations of a company's knowledge and competence, however, may usefully be guided by asking where the new areas under consideration tend to fall along the taxonomic dimensions of Figure 8–1.

There do exist important opportunities for affecting the positions that particular knowledge developments take on these dimensions. The degree of articulation of anything that is articulable is partially controllable. The possibilities for controlling observability and resisting reverse engineering are illustrated by the practice of "potting" integrated circuit devices—encasing them in a resin that cannot be removed without destroying the device (Shapley 1978). In the case of process knowledge, hazards associated with observability in use may be reduced not only by restricting observation opportunities to employees but also by compartmentalizing knowledge within the company and restricting the opportunity for a full overview to a select few. The emphasis that Teece (1986) places on control of cospecialized assets in the protection of gains from innovation is interpretable in the present framework as involving recognition that virtually any innovation is at least potentially an element of a system in one or more ways. Acquiring control of the complementary elements of the system is a way to move away from independence and its attendant hazards of involuntary transfer.

Features that restrain involuntary transfer tend to inhibit voluntary transfer; likewise, actions undertaken to facilitate voluntary transfer may well facilitate involuntary transfer also. It is here that the question of how value is to be derived becomes crucial. On the assumption that value is maximized by exploiting the knowledge within the firm (and at approximately the current scale), the general rule is to keep to the left in Figure 8–1, to the extent possible. If, on the other hand, the appropriate course is to rapidly expand the use of the same knowledge within the company or to enter licensing agreements or partnerships concerning the technological or organizational competencies involved, then it is at least necessary to restrain the tendency to keep to the left and perhaps better to keep to the right. These observations do not suggest the full complexity of the problem because actually the value-maximizing choice of mode of exploitation of the knowledge is interdependent with the cost and effectiveness of directing its development to the left or to the right. If, for example, it appears unlikely that a process secret can be kept for long, regardless of the effort put into the attempt, then either rapid growth to exploit leadtime or licensing arrangements may be a superior exploitation mode. Such a choice may imply a drastic change in how the development is handled because many arrangements that would support the secrecy approach become counterproductive. Finally, the answer to the antecedent question of whether an entire area of activity is worth entering in the first place depends on the value achievable from an appropriate simultaneous choice of position on the taxonomic dimensions and mode of exploitation of the knowledge.

Heuristic Frames

The foregoing discussion is far from exhaustive either with respect to the dimensions on which knowledge states might be described or with respect to the control variables that can be employed to alter them. Under the former heading, for example, there is obviously a need for an approach to describing the *amount* of knowledge or competence held in a particular area; under the latter, there is the whole domain of the legal system and its various mechanisms that support or hamper efforts to protect knowledge assets. The discussion is at least suggestive of a general approach to the development of heuristic frames appropriate to the development and protection of knowledge and competence assets in a particular line of business. Consider once again the example of BCG experience curve doctrine. Mechanisms that have been suggested to account for the experience curve effect include skill development by individual workers and design improvements in both products and processes. Each of these three is no doubt real, but they represent very different sorts of knowledge assets in the taxonomic scheme of Figure 8–1 and are accordingly developed and tended by quite different means (control variables).

The Diversity of Industrial Contexts

This section sets forth evidence that U.S. manufacturing industries present a diverse array of environments in terms of the characteristics of the productive knowledge on which they depend and the means that are effective in protecting knowledge assets. One obvious strategic implication of this diversity is that lessons derived from experience in one industry may be very misleading guides to knowledge-related strategic choices in another. Also, while the across-industry variation is demonstrably large, this does not imply that the within-industry variation is small. (Indeed, as just pointed out, the slope of a single experience curve may reflect a variety of mechanisms.) Consideration of the across-industry variation suggests that there may be a payoff to a more discriminating approach to the problems of a single business unit—a willingness to recognize, for example, that a particular situation arising in some branch of chemical manufacturing may be "like semiconductors" even though most of the situations encountered there are, naturally, "like chemicals."[9]

The evidence presented here derives from the Yale survey of R&D executives. This survey project, headed by Richard Levin in collaboration with Alvin Klevorick, Richard Nelson, and myself, obtained answers from R&D executives to a wide range of questions regarding the appropriability of gains from innovation and the sources of technological opportunity in the industries with which they are familiar. To circumvent problems of confidentiality regarding the practices of individual companies, the survey addressed respondents in their capacity as experts on innovation in particular lines of business, rather than as authorities on the practices of their own companies. We obtained 650 responses from executives involved in R&D management in 130 lines of business. There are eighteen lines of business, noncoincidentally including most of the major R&D performing industries, for which we received ten or more responses.[10]

The first questions of the survey asked respondents to record, on a seven-point Likert scale, their view of the effectiveness of various means of protecting the returns from innovation in their various lines of business. Table 8–1 records the exact phrasing of the question, along with mean effectiveness scores for patents to prevent duplication obtained from the eighteen industries for which we had ten or more responses.

The first notable feature of these results is that, with the single exception of petroleum refining, patents are consistently rated as more effective in protecting product innovations than process innovations. To interpret this pattern requires consideration of how the observability-in-use dimension of a knowledge asset interacts with the patent law. In general (and practices like "potting" notwithstanding), the design information embodied in a product is discoverable through reverse engineering by any purchaser who wants to apply the necessary resources. Contractual arrangements to forestall reverse engineering are effective in only a few cases involving narrow markets where all buyers are identified to the seller. Also, design information is articulable. By contrast, physical access to a production process can be restricted, and much process knowledge is tacit or unarticulated. Although there are cases in which knowledge of the process can be inferred from reverse engineering of the product, the general situation is that process knowledge need not be observable in use unless the producer permits it. Finally, there are many cases in

Table 8–1.   Effectiveness Ratings of Patents to Prevent Duplication

Question: In this line of business, how effective is each of the following means of capturing and protecting the competitive advantages of new or improved processes/products? (1) Patents to prevent competitors from duplicating the product, (2) Patents to secure royalty income, (3) Secrecy, (4) Lead time (being first with a new process/product), (5) Moving quickly down the learning curve, (6) Superior sales or service efforts. Scale: 1 = "not at all effective," 4 = "moderately effective," 7 = "very effective." (Mean scores are given for eighteen industries with ten or more respondents.)

Line of Business[a]	New Processes	New Products
Pulp, paper, and paperboard	2.58	3.33
Inorganic chemicals	4.59	5.24
Organic chemicals	4.05	6.05
Drugs	4.88	6.53
Cosmetics	2.94	4.06
Plastic materials	4.58	5.42
Plastic products	3.24	4.93
Petroleum refining	4.90	4.33
Steel mill products	3.50	5.10
Pumps and pumping equipment	3.18	4.36
Motors, generators, and controls	2.70	3.55
Computers	3.33	3.43
Communications equipment	3.13	3.65
Semiconductors	3.20	4.50
Motor vehicle parts	3.65	4.50
Aircraft and parts	3.14	3.79
Measuring devices	3.60	3.89
Medical instruments	3.20	4.73

a. See Table 8–2 for SIC codes.

which complex processes yield simple products. On all these grounds, knowledge embodied in products is inherently more subject to involuntary transfer than process knowledge—the patent system aside.

As is apparent from Table 8–1, patenting is not always an effective response to this inherent vulnerability of product information. In relation to process knowledge, however, the patent system has specific and severe disabilities. Processes that involve an inarticulable "mental step" are not patentable. Insights into process arrangements that are of such broad scope as to amount to "ideas" or "scientific principles" are not patentable. Most important, a patent on a process that is not observable in use (without consent) is difficult to enforce because infringement is not directly observable and can at best be inferred, in the first instance, from circumstantial evidence. Finally, the patent application itself discloses information that might otherwise be protected by secrecy. In sum, patent protection is ineffective for processes relative to products because tacitness and non-observability are more characteristic of process than of product innovations. In addition, secrecy is a strong alternative to patenting for processes but a weak alternative for products.

A second significant observation derivable from Table 8–1 is that highly effective patent protection is by no means a necessary condition for technological progressiveness. At least, this is the conclusion that follows if we assume that common perception is correct in assessing such industries as computers, semiconductors, communications equipment, and aircraft as being highly innovative. Of the eight mean scores characterizing patent effectiveness for processes and products in these four industries, only one (semiconductor products) is above the humdrum midlevel—"moderately effective." A thorough analysis of this phenomenon is beyond the scope of the present discussion, but it is worth noting the obvious inference that if these industries manage to be innovative in spite of the lack of effective patent protection, other mechanisms presumably protect the gains from innovation. Some reflection on the nature of these industries and another glance at Figure 8–1 produce the suggestion that the prevalence of "complexity" and the status "part of a system" may play a key role in protecting innovative gains in these areas.

A third prominent feature of Table 8–1 is the high scores for patent effectiveness shown for a group of industries that are within the SIC chemicals group (SIC 28). Although the importance of patents in this area comes as no surprise—especially in the case of drugs—the survey results are nevertheless striking for the closeness of the association between patent effectiveness and the chemical industry. In Table 8–1, SIC 28 accounts for all four out of the top four mean scores in the product column, and four out of the top five in the process column—and the intruder (with the top score) is a close relative of the chemical business, petroleum refining.[11] Out of the full sample of 130 lines of business, there are only four for which patents are rated more effective than any of the nonpatent means of protecting gains from product innovation listed in question 1B. Two of those are drugs and pesticides, one is represented by a single respondent, the other is meat products (three respondents).

At least a part of this pattern is easily accounted for by reference to the taxonomic dimensions of Figure 8–1. Chemical products are plainly far to the right on the dimensions of articulability, observability in use, and independence (meaning in this case that there are markets for individual products as defined by the molecular structure). Indeed, with modern analytical techniques, chemical products (whether simple or complex) are essentially an open book to the qualified and well-equipped observer. In the absence of patents, this sort of knowledge would be highly subject to involuntary transfer. On the other hand, essentially the same characteristics make the property right conferred by a chemical product patent peculiarly sharply defined, distinctively invulnerable to "inventing around." (As is documented elsewhere in the survey results, "inventing around" is the leading reason given for ineffectiveness of patent protection.)

The rationale for the distinctive position of chemicals industries with respect to patent protection of new processes seems much less clear. Relatively tight process-product links may well be involved; for example, it may be unusually easy to infer process patent infringement from examination of the product.

A different perspective on the diversity of industrial contexts is provided by Table 8–2. This shows, for the same industries as Table 8–1, the method of protecting gains from process and product innovation that ranked highest in mean score. Note that one of the possible responses, patents to secure royalty income, never shows up in first place.[12] Patents to prevent duplication rank first only in drugs and in organic chemical

Table 8–2. Highest-Rated Methods of Protecting Gains
from Innovation

Key: PD = patents to prevent competitors from duplicating the product, S = secrecy, LT = lead time, LC = moving quickly down the learning curve, SS = superior sales or service efforts. Numbers in parentheses are SIC codes for the line of business. (Industries and question are the same as in Table 8–1.)

Line of Business	Processes	Products
Pulp, paper, and paperboard (261, 262, 263)	LC	SS
Inorganic chemicals (2812, 2819)	LT	LT
Organic chemicals (286)	S	PD
Drugs (283)	PD	PD
Cosmetics (2844)	LT	LT
Plastic materials (2821)	LT	SS
Plastic products (307)	LC	SS
Petroleum refining (291)	LC	SS
Steel mill products (331)	LC	SS
Pumps and pumping equipment (3561)	LT	LT
Motors, generators, and controls (3621, 3622)	LT	SS
Computers (3573)	LT	LT
Communications equipment (3661, 3662)	LT	SS
Semiconductors (3674)	LC	LT
Motor vehicle parts (3714)	LC	LT
Aircraft and parts (3721, 3728)	LT	LT
Measuring devices (382)	LT-LC	SS
Medical instruments (3841, 3842)	LT	SS

products. Surprisingly, only in organic chemicals does secrecy hold first place as a means of protecting process innovation. Everywhere else, it is lead time or learning curve effects in the process column and lead time or sales and service efforts in the product column.

Finally, it should be noted explicitly that the industries in which we obtained ten or more responses from R&D executives are industries that are doing a substantial amount of R&D—and some mechanism must be protecting the results of that effort or it would not be going on. There is another industrial environment, not represented in Tables 8–1 and 8–2, where appropriability is low and not much R&D is done. This does not mean that knowledge and competence are strategically insignificant. It means that emphasis shifts from the production of innovation to the adoption of innovations produced elsewhere and to the competent use of competitive weapons other than innovation.

Summary

Considering the acknowledged importance of knowledge and competence in business strategy and indeed in the entire system of contemporary human society, it is striking that there seems to be a paucity of language useful for discussing the subject. Within each microcosm of expertise or skill, there is of course a specialized language in which *that* subject—or at least the articulable parts of it—can be discussed. At the opposite extreme, there is terminology of very broad scope. There are words like *information, innovation, skill, technology transfer, diffusion, learning,* and (of course) *knowledge* and *competence.* These name parts of the realm of discourse but do not do much to point the way toward advancing the discourse. The problems of managing technological and organizational change surely lie between these two extremes of low and high generality, and in that range there seems to be a serious dearth of appropriate terminology and conceptual schemes. Such, at least, has been the premise of this paper.

Like evolutionary theory, from which it partly derives, the approach to strategy sketched in earlier sections has a healthy appetite for facts. The first implementation challenge it poses is that of strategic state description—the development of a strategically useful characterization of those features of the organization that are not subject to choice in the short run but are influencable in the long run and are considered key to its development and success. From evolutionary theory comes the idea that a state description may include organizational behavioral patterns or routines that are not amenable to rapid change, as well as a more conventionally defined assets. It is by this route that a variety of consid-erations that fall under the rubrics *knowledge* and *competence* may enter the strategic state description. From the optimal control theory side comes, first, the insistence that a scheme intended to provide policy guidance must have some choices to relate to; there must be some control variables that affect the development of the state variables. Second, granting the adoption of present value or expected present value as a criterion, control theory provides an approach to the valuation of strategic assets, conventional, and unconventional. In particular, it offers the full imputation challenge of seeking out the present acorns from which the future golden oaks are expected to grow—information likely to be useful in the proper tending of the acorns.

Both control theory and evolutionary theory invoke the notion of state description in the context of formal modeling, and both pursue conclusions logically within the context of the model once it is set down. Without denying the possible usefulness of a formal modeling approach to particular strategic problems, I have proposed the informal, looser and more flexible concept of a heuristic frame—essentially, the control theory approach stripped down to a list of state descriptors and controls. The word *heuristic* serves as a reminder that a control theory model or any other elaboration of a heuristic frame is an elaboration of an educated guess about what matters and what can be controlled. The making of that guess may be the key step. A change of heuristic frame may make all the difference in terms of identifying those acorns from which future value will grow and in so doing dramatically affect the value actually realized. It is in the choice of heuristic frame, above all, that creative insight into strategic problems plays its role.

Another challenging problem is the subtask of the strategic state description task that involves the characterization of knowledge and competence assets. Such assets are extraordinarily diverse, not only (obviously) in their specific details but also in a number of identifiable dimensions of strategic significance. Some assets are subject to major hazards of involuntary transfer; others may prove highly resistant to affirmative efforts to transfer them pursuant to a sale or exchange. Among the control variables involved in the management of these assets are some that affect the hazards of involuntary transfer or the feasibility of voluntary transfer. It was suggested that the opposition between these two control goals is fundamental.

Finally, the diversity of knowledge contexts found in U.S. manufacturing industry (documented with evidence from the Yale survey of R&D executives) can be analyzed using the chapter's taxonomic scheme. It underscores the point that different approaches to the management of knowledge and competence assets, for understandable reasons, predominate in different industries.

The themes introduced in this chapter are not fully developed. There is much more to be done and said. If the gaps and loose ends are obvious, that at least suggests that a heuristic frame for the analysis of knowledge and competence assets has been put in place.

NOTES

1. The habits may well be sophisticated skills and the impulses leaps of passionate faith. Peters, in the paper cited, puts forward normative conclusions that emphasize the role of corporate leaders in the related tasks of (1) selecting, in an uncertain world, directions of skill development that will lead to long-term profitability and (2) guiding the development of these skills through the enunciation and legitimation of appropriate subgoals. He seems to be skeptical of the usefulness of any (intendedly) rational analytic framework in connection with step 1.

2. There is a well-known device for extending the range of goals represented in an optimization problem beyond that reflected by the formal criterion of the problem: Introduce constraints requiring that acceptable levels of other goal variables be achieved (see Simon 1964). This device is one available method for extending the propositions developed here. For example, "Maximize present value subject to acting in compliance with the law" is formally a close cousin of "Maximize present value," although it may be a very different thing substantively.

3. In one way or another, most contemporary theorists actively concerned with the theory of the firm and its internal organization seem to have adopted the nexus-of-contracts view. For a concise statement of this view, see the opening paragraph of Fama and Jensen (1985).

4. As a classroom example of this problem, I like to cite the following passage from a BCG publication, regarding the appropriate treatment of "dogs": "They can never be satisfactorily profitable and should be liquidated in as clever and graceful a manner as possible. *Outright sale to a buyer with different perceptions can sometimes be accomplished*" (Conley 1970: 13, emphasis supplied). This is sound advice—but, of course, equally sound for stars, cash cows, and question marks as for dogs, if the buyer's perceptions differ sufficiently in the right direction. On the other hand, if no such buyer can be found for a dog, it is not at all clear that selling it makes sense.

5. In economic theory, the issue and principle of full imputation may be traced back at least to Wicksteed's concern with the exhaustion-of-the-product problem in production theory (Wicksteed 1894). For a sophisticated discussion of profitability and imputation, see Triffin (1949: ch. 5).

6. Failure can be explained by deficiencies in these various respects—although luck seems to be more widely acknowledged as an explanation for failure than for success.

7. A noteworthy example of the latter genre is Salinger (1984), which gives impressive evidence that an R&D stock (accumulated expenditure) variable is an important determinant of profitability when the latter is measured by Tobin's q. This approach to assessing profitability is in much better conformity with the valuation discussion in this chapter than approaches based exclusively on accounting measures.

8. The term *fissioning* is used by Charles A. Ziegler (1985: 104), who provides an interesting account of the process, concluding that "it can be unwise for corporate leaders to overestimate the efficacy of legal precautions and countermeasures and to underestimate the debilitating effect on the parent firm that successful fissioning can produce."

9. There are analogous implications for public policy. For example, in the light of the evidence presented below, it is clear that patent policy is a tool of dramatically different significance in different industries. It is, accordingly, an inappropriate focus

for policy attention arising from general (as opposed to industry-specific) concerns about the innovativeness of U.S. industry (see Levin 1985).

10. The lines of business were those defined by the Federal Trade Commission, which correspond primarily to four-digit SIC manufacturing industries, occasionally to groups of four-digit industries or to three-digit industries. Lines of business that report zero R&D expenditures were excluded from the sample, as were highly heterogeneous ones. The starting point for our list of R&D performing firms was the *Business Week* annual R&D survey list, which includes all publicly traded firms that report R&D expenses in excess of 1 percent of sales or $35 million. This gives the survey excellent coverage as measured by R&D expenditure but largely omits the perspective of small, entrepreneurial firms. For further details on the design of the survey and an overview of its results, see Levin, *et al.* (1984).

11. The anomalies relative to this relationship are cosmetics (actually "perfumes, cosmetics and toilet preparations") (an industry which does not seem to share the characteristics of the chemicals club in spite of being SIC 2844) and steel mill products (SIC 331), which score high on patent effectiveness for new products in spite of being remote from the chemicals club.

12. By contrast, the analysis of situations in which patents are used to secure royalty income does occupy a very prominent place in the theoretical economics of innovation.

REFERENCES

Bellman, R. 1957. *Dynamic Programming.* Princeton, N.J.: Princeton University Press.

Blakeslee, S. 1985. "Clues Hint at Brain's Two Memory Maps." *New York Times* (February 19): C1.

Conley, P. 1970. *Experience Curves as a Planning Tool.* Boston: Boston Consulting Group.

Cyert, R.M., and J.G. March. 1963. *A Behavioral Theory of the Firm.* Englewood Cliffs, N.J.: Prentice-Hall.

Dorfman, R., P.A. Samuelson, and R.M. Solow. 1958. *Linear Programming and Economic Analysis.* New York: McGraw-Hill.

Fama, E.F., and M.C. Jensen. 1985. "Organizational Form and Investment Decisions." *Journal of Financial Economics* 14: 101–19.

Flax, S. 1984. "How to Snoop on Your Competitors." *Fortune* (May 14): 29–33.

Greiner, L.E. 1983. "Senior Executives as Strategic Actors." *New Management* 1: 11–15.

Levin, R. 1985. "Patents in Perspective." *Antitrust Law Journal* 53: 519–22.

Levin, R., A.K. Klevorick, R.R. Nelson, and S.G. Winter. 1984. "Survey Research on R&D Appropriability and Technological Opportunity." Working paper, Yale University.

Nelson, R.R., and S.G. Winter. 1982. *An Evolutionary Theory of Economic Change.* Cambridge, Mass.: Belknap Press of the Harvard University Press.

Peters, T.J. 1984. "Strategy Follows Structure: Developing Distinctive Skills." In *Strategy and Organization: A West Coast Perspective*, edited by G. Carroll and D. Vogel. Mansfield, Mass.: Pitman.

Polanyi, M. 1962. *Personal Knowledge: Toward a Post-Critical Philosophy.* New York: Harper Torchbooks.

Pontryagin, L.S., V.G. Boltyanskii, R.V. Gamkrelidze, and E.F. Mischenko. 1962. *The Mathematical Theory of Optimal Processes.* New York: Interscience.

Porter, M.E. 1980. *Competitive Strategy: Techniques for Analyzing Industries and Competitors.* New York: Free Press.

Salinger, M.A. 1984. "Tobin's q Unionization, and the Concentration-Profits Relationship." *The Rand Journal of Economics* 15: 159–70.

Shapley, D. 1978. "Electronics Industry Takes to 'Potting' Its Products for Market." *Science* 202, 848–49.

Simon, H.A. 1957. *Administrative Behavior*, 2d ed. New York: Free Press.

———. 1964. "On the Concept of Organizational Goal." *Administrative Science Quarterly* 9: 1–22.

Teece, D.J. 1986. "Capturing Value from Technological Innovation: Integration, Strategic Partnering and Licensing Decisions." Working paper, University of California, Berkeley.

Triffin, R. 1949. *Monopolistic Competition and General Equilibrium Theory*. Cambridge, Mass.: Harvard University Press.

Wicksteed, P. 1894. *An Essay on the Coordination of the Laws of Distribution*. London: Macmillan.

Ziegler, C.A. 1985. "Innovation and the Imitative Entrepreneur." *Journal of Economic Behavior and Organization* 6: 103–21.

9 PROFITING FROM TECHNO-LOGICAL INNOVATION: IMPLICATIONS FOR INTEGRA-TION, COLLABORATION, LICENSING AND PUBLIC POLICY

David J. Teece

It is quite common for innovators—those firms that are first to commercialize a new product or process in the market—to lament the fact that competitors and imitators have profited more from the innovation than the firm first to commercialize it! Because it is often held that being first to market is a source of strategic advantage, the clear existence and persistence of this phenomenon may appear perplexing if not troubling. The aim of this chapter is to explain why a fast second or even a slow third might outperform the innovator. The message is particularly pertinent to those science- and engineering-driven companies that labor under the mistaken illusion that developing new products that meet customer needs will ensure fabulous success. It may possibly do so for the product but not for the innovator.

I thank Raphael Amit, Harvey Brooks, Chris Chapin, Therese Flaherty, Richard Gilbert, Heather Haveman, Mel Horwitch, David Hulbert, Carl Jacobsen, Michael Porter, Gary Pisano, Richard Rumelt, Raymond Vernon, and Sidney Winter for helpful discussions relating to the subject matter of this chapter. Three anonymous referees also provided valuable criticisms. I gratefully acknowledge the financial support of the National Science Foundation under grant no. SRS-8410556 to the Center for Research in Management, University of California, Berkeley. Earlier versions of this chapter were presented at a National Academy of Engineering Symposium titled "World Technologies and National Sovereignty," February 1986, and at a conference on innovation at the University of Venice, March 1986.

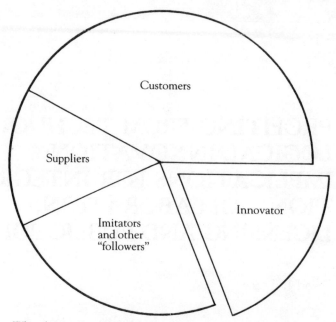

What determines the share of profits captured by the innovator?

Figure 9–1. Explaining the Distribution of the Profits from Innovation.

In this chapter, a framework is offered that identifies the factors that determine who wins from innovation: the firm that is first to market, follower firms, or firms that have related capabilities that the innovator needs. The follower firms may or may not be imitators in the narrow sense of the term, although they sometimes are. The framework appears to have utility for explaining the share of the profits from innovation accruing to the innovator compared to its followers and suppliers (see Figure 9–1), as well as for explaining a variety of interfirm activities such as joint ventures, coproduction agreements, cross-distribution arrangements, and technology licensing. Implications for strategic management, public policy, and international trade and investment are also discussed.

The Phenomenon

Figure 9–2 presents a simplified taxonomy of the possible outcomes from innovation. Quadrant 1 represents positive outcomes for the innovator. A first-to-market advantage is translated into a sustained competitive advantage that either creates a new earnings stream or enhances an existing one. Quadrant 4 and its corollary quadrant 2 are the ones that are the focus of this paper.

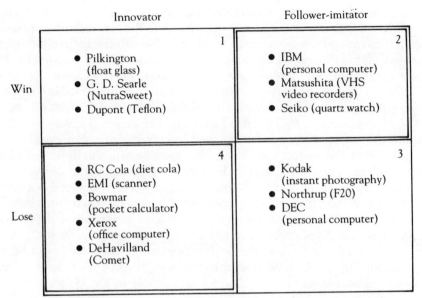

Figure 9–2.   Taxonomy of Outcomes from the Innovation Process.

The EMI CAT Scanner is a classic case of the phenomenon to be investigated.[1] By the early 1970s, the U.K. firm Electrical Musical Industries (EMI) Ltd. was in a variety of product lines including phonographic records, movies, and advanced electronics. EMI had developed high resolution televisions in the 1930s, pioneered airborne radar during World War II, and developed the U.K.'s first all-solid-state computers in 1952.

In the late 1960s Godfrey Hounsfield, an EMI senior research engineer engaged in pattern recognition research that resulted in his displaying a scan of a pig's brain. Subsequent clinical work established that computerized axial tomography (CAT) was viable for generating cross-sectional views of the human body, the greatest advance in radiology since the discovery of x-rays in 1895.

Although EMI was initially successful with its CAT scanner, within six years of its introduction into the United States in 1973 the company had lost market leadership and by the eighth year had dropped out of the CT scanner business. Other companies successfully dominated the market, although they were late entrants and are still profiting in the business today.

Other examples include RC Cola, a small beverage company that was the first to introduce cola in a can and the first to introduce diet cola. Both Coca Cola and Pepsi followed almost immediately and deprived RC of any significant advantage from its innovation. Bowmar, which introduced the pocket calculator, was not able to withstand competition from Texas Instruments, Hewlett Packard, and others and went out of business. Xerox failed to succeed with its entry into the office computer business, even though Apple succeeded with the MacIntosh, which contained many of Xerox's key product ideas, such as the mouse and icons. The de Havilland Comet saga has some of the same features. The Comet I jet was introduced into the commercial-airline business two years or so before Boeing introduced the 707, but de Havilland failed to capitalize on its substantial early advantage. MITS introduced the first personal computer, the Altair, experienced a burst of sales, then slid quietly into oblivion.

If there are innovators who lose, there must be follower imitators who win. A classic example is IBM with its PC, a great success since the time it was introduced in 1981. Neither the architecture nor components embedded in the IBM PC were considered advanced when introduced, nor was the way the technology was packaged a significant departure from then-current practice. Yet the IBM PC was fabulously successful and established MS-DOS as the leading operating system for 16-bit PCs. By the end of 1984, IBM had shipped over 500,000 PCs, and many considered that it had irreversibly eclipsed Apple in the PC industry.

Profiting from Innovation: Basic Building Blocks

In order to develop a coherent framework within which to explain the distribution of outcomes illustrated in Figure 9–2, three fundamental building blocks must first be put in place: the appropriability regime, the dominant design paradigm, and complementary assets.

Regimes of Appropriability

A regime of appropriability refers to the environmental factors, excluding firm and market structure, that govern an innovator's ability to capture the profits generated by an innovation. The most important dimensions of such a regime are the nature of the technology and the efficacy of legal mechanisms of protection (Figure 9–3).

It has long been known that patents do not work in practice as they do in theory. Rarely, if ever, do patents confer perfect appropriability although they do afford considerable protection on new chemical products and rather simple mechanical inventions. Many patents can be "invented around" at modest costs. They are especially ineffective at protecting process innovations. Often patents provide little protection because the legal requirements for upholding their validity or for proving their infringement are high.

In some industries, particularly where the innovation is embedded in processes, trade secrets are a viable alternative to patents. Trade-secret

- Legal instruments
  - Patents
  - Copyrights
  - Trade secrets

- Nature of technology
  - Product
  - Process
  - Tacit
  - Codified

Figure 9–3.   Appropriability Regime: Key Dimensions.

protection is possible, however, only if a firm can put its product before the public and still keep the underlying technology secret. Usually only chemical formulas and industrial-commercial processes (such as cosmetics and recipes) can be protected as trade secrets after they are out.

The degree to which knowledge is tacit or codified also affects ease of imitation. Codified knowledge is easier to transmit and receive and is more exposed to industrial espionage and the like. Tacit knowledge by definition is difficult to articulate, so transfer is hard unless those who possess the know-how in question can demonstrate it to others (Teece 1981). Survey research indicates that methods of appropriability vary markedly across industries and probably within industries as well (Levin *et al.* 1984).

The property-rights environment within which a firm operates can thus be classified according to the nature of the technology and the efficacy of the legal system to assign and protect intellectual property. Although a gross simplification, a dichotomy can be drawn between environments in which the appropriability regime is tight (technology is relatively easy to protect) and weak (technology is almost impossible to protect). Examples of the former include the formula for Coca Cola syrup; an example of the latter would be the Simplex algorithm in linear programming.

The Dominant Design Paradigm

It is commonly recognized that there are two stages in the evolutionary development of a given branch of a science: the preparadigmatic stage, when there is no single generally accepted conceptual treatment of the phenomenon in a field of study, and the paradigmatic stage, which begins when a body of theory appears to have passed the canons of scientific acceptability. The emergence of a dominant paradigm signals scientific maturity and the acceptance of agreed on standards by which what has been referred to as normal scientific research can proceed. These standards remain in force unless or until the paradigm is overturned. Revolutionary science is what overturns normal science, as when Copernicus's theories of astronomy overturned Ptolemy's in the seventeenth century.

Abernathy and Utterback (1978) and Dosi (1982) have provided a treatment of the technological evolution of an industry that appears to parallel Kuhnian notions of scientific evolution (see Kuhn 1970). In the early stages of industry development, product designs are fluid, manufacturing processes are loosely and adaptively organized, and generalized capital is used in production. Competition among firms manifests itself in competition among designs that are markedly different from each other. This might be called the preparadigmatic stage of an industry.

At some point in time and after considerable trial and error in the marketplace, one design or a narrow class of designs begins to emerge as the more promising. Such a design must be able to meet a whole set of user needs in a relatively complete fashion. The Model T Ford, the IBM 360, and the Douglas DC-3 are examples of dominant designs in the automobile, computer, and aircraft industry, respectively.

Once a dominant design emerges, competition shifts to price and away from design. Competitive success then shifts to a whole new set of variables. Scale and learning become much more important, and specialized capital gets deployed as incumbents seek to lower unit costs through exploiting economies of scale and learning. Reduced uncertainty over product design provides an opportunity to amortize specialized long-lived investments.

Innovation is not necessarily halted once the dominant design emerges; as Clark (1985) points out, it can occur lower down in the design hierarchy. For instance, a V cylinder configuration emerged in automobile engine blocks during the 1930s with the Ford V-8 engine. Niches were quickly found for it. Moreover, once the product design stabilizes, there is likely to be a surge of process innovation as producers attempt to lower production costs for the new product (see Figure 9–4).

The Abernathy-Utterback framework does not characterize all industries. It seems more suited to mass markets where consumer tastes are relatively homogeneous. It would appear to be less characteristic of small-niche markets where the absence of scale and learning economies attaches much less of a penalty to multiple designs. In these instances, generalized equipment will be employed in production.

The existence of a dominant design watershed is of great significance to the distribution of profits between innovator and follower. The innovator may have been responsible for the fundamental scientific breakthroughs as well as the basic design of the new product. However, if imitation is relatively easy, imitators may enter the fray, modifying the product in important ways yet relying on the fundamental designs pioneered by the innovator. When the game of musical chairs stops and a dominant design emerges, the innovator may well end up positioned disadvantageously relative to a follower. Hence, when imitation is pos-

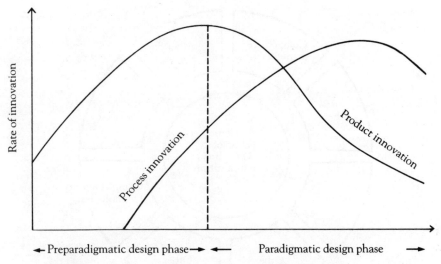

Figure 9–4.   Innovation over the Product/Industry Life Cycle.

sible and occurs coupled with design modification before the emergence of a dominant design, followers have a good chance of having their modified product anointed as the industry standard, often to the great disadvantage of the innovator.

Complementary Assets

Let the unit of analysis be an innovation. An innovation consists of certain technical knowledge about how to do things better than with the existing state of the art. Assume that the know-how in question is partly codified and partly tacit. For such know-how to generate profits, it must be sold or utilized in some fashion in the market.

In almost all cases, the successful commercialization of an innovation requires that the know-how in question be utilized in conjunction with other capabilities or assets. Services such as marketing, competitive manufacturing, and after-sales support are almost always needed. These services are often obtained from complementary assets that are specialized. For example, the commercialization of a new drug is likely to require the dissemination of information over a specialized information channel. In some cases, as when the innovation is systemic, the complementary assets may be other parts of a system. For instance, computer hardware typically requires specialized software, both for the operating system and for applications. Even when an innovation is autonomous, as with plug-compatible components, certain complementary capabilities or assets will be

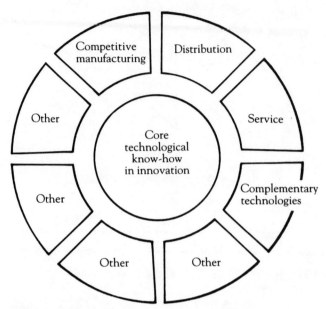

Figure 9–5. Complementary Assets Needed to Commercialize an Innovation.

needed for successful commercialization. Figure 9–5 summarizes this schematically.

Whether the assets required for least-cost production and distribution are specialized to the innovation turns out to be important in the development presented below. Accordingly, the nature of complementary assets is explained in some detail. Figure 9–6 differentiates among complementary assets that are generic, specialized, and cospecialized.

Generic assets are general-purpose assets that do not need to be tailored to the innovation in question. Specialized assets are those with unilateral dependence between the innovation and the complementary asset. Cospecialized assets are those with bilateral dependence. For instance, specialized repair facilities were needed to support the introduction of the rotary engine by Mazda. These assets are cospecialized because of the mutual dependence of the innovation on the repair facility. Containerization similarly required the deployment of some cospecialized assets in ocean shipping and terminals. However, the dependence of trucking on containerized shipping was less than that of containerized shipping on trucking as trucks can convert from containers to flat beds at low cost. An example of a generic asset would be the manufacturing

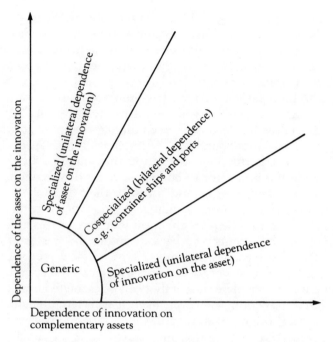

Figure 9–6. Complementary Assets: Generic, Specialized, and Cospecialized.

facilities needed to make running shoes. Generalized equipment can be employed in the main, exceptions being the molds for the soles.

Implications for Profitability

These three concepts can now be related in a way that will shed light on the imitation process and the distribution of profits between innovator and follower. We begin by examining tight appropriability regimes.

Tight Appropriability Regimes

In those few instances where the innovator has an ironclad patent or copyright protection or where the nature of the product is such that trade secrets effectively deny imitators access to the relevant knowledge, the innovator is almost assured of translating its innovation into market value for some period of time. Even if the innovator does not possess the desirable endowment of complementary costs, ironclad protection of intellectual property will afford the innovator the time to access these assets. If these assets are generic, contractual relation may well suffice;

and the innovator may simply license its technology. Specialized R&D firms are viable in such an environment. Universal Oil Products, an R&D firm developing refining processes for the petroleum industry, was one such case in point. If, however, the complementary assets are specialized or cospecialized, contractual relationships are exposed to hazards because one or both parties will have to commit capital to certain irreversible investments that will be valueless if the relationship between innovator and licensee breaks down. Accordingly, the innovator may find it prudent to expand its boundaries by integrating into specialized and cospecialized assets. Fortunately, the factors that make for difficult imitation will enable the innovator to build or acquire those complementary assets without competing with innovators for their control.

Competition from imitators is muted in this type of regime, which sometimes characterizes the petrochemical industry. In this industry, the protection offered by patents is fairly easily enforced. One factor assisting the licensee in this regard is that most petrochemical processes are designed around a specific variety of catalysts which can be kept proprietory. An agreement not to analyze the catalyst can be extracted from licensees, affording extra protection. However, even if such requirements are violated by licensees, the innovator is still well positioned because the most important properties of a catalyst are related to its physical structure, and the process for generating this structure cannot be deduced from structural analysis alone. Every reaction technology a company acquires is thus accompanied by an ongoing dependence on the innovating company for the catalyst appropriate to the plant design. Failure to comply with various elements of the licensing contract can thus result in a cutoff in the supply of the catalyst and possibly in facility closure.

Similarly, if the innovator comes to market in the preparadigmatic phase with a sound product concept but the wrong design, a tight appropriability regime will afford the innovator the time needed to perform the trials needed to get the design right. As discussed earlier, the best initial design concepts often turn out to be hopelessly wrong, but if the innovator possesses an impenetrable thicket of patents or has technology that is simply difficult to copy, then the market may well afford the innovator the necessary time to ascertain the right design before being eclipsed by imitators.

Weak Appropriability

Tight appropriability is the exception rather than the rule. Accordingly, innovators must turn to business strategy if they are to keep imitator followers at bay. The nature of the competitive process will vary according to whether the industry is in the paradigmatic or the preparadigmatic phase.

Preparadigmatic Phase. In the preparadigmatic phase, the innovator must be careful to let the basic design float until sufficient evidence has accumulated that a design has been delivered that is likely to become the industry standard. In some industries, there may be little opportunity for product modification. In microelectronics, for example, designs become locked in when the circuitry is chosen. Product modification is limited to debugging and software modification. An innovator must begin the design process anew if the product does not fit the market well. In some respects, however, selecting designs is dictated by the need to meet certain compatibility standards so that new hardware can interface with existing applications software. In one sense, therefore, the design issue for the microprocessor industry today is relatively straightforward: deliver greater power and speed while meeting the computer industry standards of the existing software base. However, from time to time, windows of opportunity emerge for the introduction of entirely new families of microprocessors that will define a new industry and software standard. In these instances, basic design parameters are less well defined and can be permitted to float until market acceptance is apparent.

The early history of the automobile industry exemplifies exceedingly well the importance for subsequent success of selecting the right design in the preparadigmatic stages. None of the early producers of steam cars survived the early shakeout when the closed-body internal-combustion-engine automobile emerged as the dominant design. The steam car, nevertheless, had numerous early virtues such as reliability, which the internal-combustion-engine autos could not deliver.

The British fiasco with the Comet I is also instructive. De Havilland had picked an early design with both technical and commercial flaws. By moving into production, de Havilland was hobbled by significant irreversibilities and loss of reputation to such a degree that it was unable to convert to the Boeing design, which subsequently emerged as dominant. It was not able to occupy even second place, which went instead to Douglas.

As a general principle, it appears that innovators in weak appropriability regimes need to be coupled intimately to the market so that user needs can fully influence designs. When multiple parallel and sequential prototyping is feasible, it has clear advantages. Generally such an approach is prohibitively costly. When development costs for a large commercial aircraft exceed $1 billion, variations on a theme are all that is possible.

Hence, the probability that an innovator—defined here as a firm that is first to commercialize a new product design concept—will enter the paradigmatic phase possessing the dominant design is problematic. The lower the cost of prototyping, the higher the relative probabilities and the more tightly coupled the firm is to the market. The latter is a function of organizational design and can be influenced by managerial choices.

The cost is embedded in the technology and cannot be influenced except in minor ways by managerial decisions. In industries with large developmental and prototyping costs—and hence significant irreversibilities—where innovation of the product concept is easy, one would expect that the probability that the innovator would emerge as the winner or among the winners at the end of the preparadigmatic stage is low.

Paradigmatic Stage. In the preparadigmatic phase, complementary assets do not loom large. Rivalry is focused on trying to identify the design that will be dominant. Production volumes are low, and there is little to be gained in deploying specialized assets as scale economies are unavailable and price is not a principal competitive factor. However, as the leading design or designs begin to be revealed by the market, volumes increase and opportunities for economies of scale will induce firms to begin gearing up for mass production by acquiring specialized tooling and equipment and possibly specialized distribution as well. Because these investments involve significant irreversibilities, producers are likely to proceed with caution. Islands of specialized capital will begin to appear in an industry that otherwise features a sea of general-purpose manufacturing equipment.

However, as the terms of competition begin to change and prices become increasingly unimportant, access to complementary assets becomes absolutely critical. Because the core technology is easy to imitate, by assumption, commercial success swings on the terms and conditions on which the required complementary assets can be accessed.

It is at this point that specialized and cospecialized assets become critically important. Generalized equipment and skills, almost by definition, are always available in an industry, and even if they are not, they do not involve significant irreversibilities. Accordingly, firms have easy access to this type of capital, and even if there is insufficient capacity available in the relevant assets, it can be put in place easily as it involves few risks. Specialized assets, on the other hand, involve significant irreversibilities and cannot be easily accessed by contract as the risks are significant for the party making the dedicated investment. The firms that control the cospecialized assets, such as distribution channels, specialized manufacturing capacity, and so forth are clearly advantageously positioned relative to an innovator. Indeed, in rare instances where incumbent firms possess an airtight monopoly over specialized assets and the innovator is in a regime of weak appropriability, all of the profits to the innovation could conceivably inure to the firms possessing the specialized assets, who should be able to get the upperhand.

Even when the innovator is not confronted by situations where competitors or potential competitors control key assets, the innovator may still be disadvantaged. For instance, the technology embedded in cardiac pacemakers was easy to imitate, and so competitive outcomes quickly came to be determined by who had easiest access to the complementary assets, in this case specialized marketing. A similar situation has recently arisen in the United States with respect to personal computers. As an industry participant (Norman 1986: 438) recently observed:

> There are a huge number of computer manufacturers, companies that make peripherals (e.g., printers, hard disk drives, floppy disk drives), and software companies. They are all trying to get marketing distributors because they cannot afford to call on all of the U.S. companies directly. They need to go through retail distribution channels, such as Businessland, in order to reach the marketplace. The problem today, however, is that many of these companies are not able to get shelf space and thus are having a very difficult time marketing their products. The point of distribution is where the profit and the power are in the marketplace today.

Channel Strategy Issues

The above analysis indicates how access to complementary assets, such as manufacturing and distribution, on competitive teams is critical if the innovator is to avoid handing over the lion's share of the profits to imitators, and/or to the owners of the complementary assets that are specialized or cospecialized to the innovation. It is now necessary to delve more deeply into the appropriate control structure that the innovator ideally ought to establish over these critical assets.

There are a myriad of possible channels that could be employed. At one extreme, the innovator could integrate into all of the necessary complementary assets, as illustrated in Figure 9–7, or into just some of them, as illustrated in Figure 9–8. Complete integration (Figure 9–7) is likely to be unnecessary as well as prohibitively expensive. It is well to recognize that the variety of assets and competences that need to be accessed is likely to be quite large even for only modestly complex technologies. To produce a personal computer, for instance, a company needs access to expertise in semiconductor technology, display technology, disk-drive technology, networking technology, keyboard technology, and several others. No company by itself can keep pace in all of these areas.

At the other extreme, the innovator could attempt to access these assets through straightforward contractual relationships (such as component supply contracts, fabrication contracts, and service contracts). In many instances, such contracts may suffice, although sometimes exposing the innovator to various hazards and dependencies that it may well wish to avoid. Between the fully integrated and fully contractual extremes, there are a myriad of intermediate forms and channels available. An analysis of the properties of the two extreme forms is presented below. A brief synopsis of mixed modes then follows.

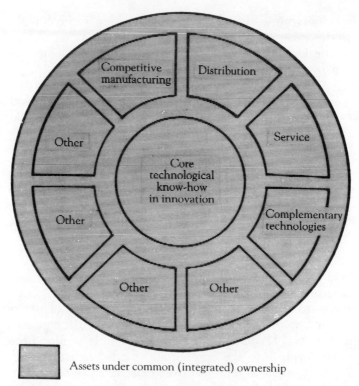

Assets under common (integrated) ownership

Figure 9–7. Complementary Assets Internalized for Innovation: Hypothetical Case #1 (Innovator Integrated into All Complementary Assets).

Contractual Modes

The advantages of a contractual solution—whereby the innovator signs a contract, such as a license, with independent suppliers, manufacturers, or distributors—are obvious. The innovator will not have to make the upfront capital expenditures needed to build or buy the assets in question, thus reducing risks as well as cash requirements.

Contracting, rather than integrating, is likely to be the optimal strategy when the innovator's appropriability regime is tight and the complementary assets are available in competitive supply (that is, there is adequate capacity and a choice of sources).

Both conditions apply in petrochemicals—for instance, so that an innovator does not need to be integrated to be successful. Consider first the appropriability regime. As discussed earlier, the protection offered by patents is fairly easily enforced, particularly for process technology, in the petrochemical industry. Given the advantageous feedstock prices available

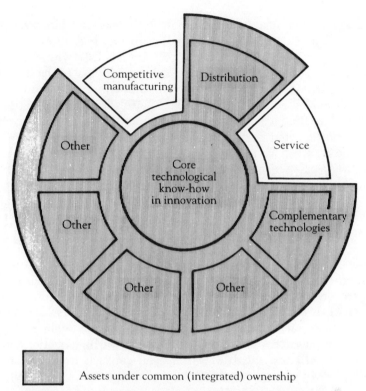

Figure 9–8. Complementary Assets Internalized for Innovation: Hypothetical Case #2 (Innovator Subcontracts for Manufacturing and Service).

in hydrocarbon rich petrochemical exporters and the appropriability regime characteristic of this industry, there is no incentive or advantage in owning the complementary assets (production facilities) as they are not typically highly specialized to the innovation. Union Carbide appears to realize this and has recently adjusted its strategy accordingly. Essentially, Carbide is placing its existing technology into a new subsidiary, Engineering and Hydrocarbons Service. The company is engaging in licensing and offers engineering, construction, and management services to customers who want to take their feedstocks and integrate them forward into petrochemicals. But Carbide itself appears to be backing away from an integration strategy.

Chemical and petrochemical product innovations are not quite so easy to protect, thus raising new challenges to innovating firms in the developed nations as they attempt to shift out of commodity petrochemicals. There are already numerous examples of new products that made it to the marketplace, filled a customer need, but, because of imitation, never generated competitive returns to the innovator. For example, in the 1960s, Dow decided to start manufacturing rigid polyurethane foam. However, it was imitated very quickly by numerous small firms that had lower costs.[2] The absence of low-cost manufacturing capability left Dow vulnerable.

Contractual relationships can bring added credibility to the innovator, especially if the innovator is relatively unknown when the contractual partner is established and viable. Indeed, arm's-length contracting that embodies more than a simple buy/sell agreement is becoming so common and is so multifaceted that the term *strategic partnering* has been devised to describe it. Even large companies such as IBM are now engaging in it. For IBM, partnering buys access to new technologies enabling the company to "learn things we couldn't have learned without many years of trial and error."[3] IBM's arrangement with Microsoft to use the latter's MS-DOS operating-system software on the IBM PC facilitated the timely introduction of IBM's personal computer into the market.

Smaller, less integrated companies are often eager to sign on with established companies because of name recognition and reputation spill-overs. For instance, Cipher Data Products, Inc. contracted with IBM to develop a low-priced version of IBM's 3480 .5-inch streaming cartridge drive, which is likely to become the industry standard. As Cipher management points out, "one of the biggest advantages to dealing with IBM is that, once you've created a product that meets the high quality standards necessary to sell into the IBM world, you can sell into any arena."[4] Similarly, IBM's contract with Microsoft "meant instant credibility" to Microsoft (McKenna 1985: 94).

It is most important to recognize, however, that strategic (contractual) partnering, which is currently very fashionable, is exposed to certain hazards, particularly for the innovator, when the innovator is trying to use contracts to access specialized capabilities. First, it may be difficult to induce suppliers to make costly irreversible commitments that depend for their success on the success of the innovation. To expect suppliers, manufacturers, and distributors to do so is to invite them to take risks along with the innovator. The problem that this poses for the innovator is similar to the problems associated with attracting venture capital. The innovator must persuade its prospective partner that the risk is a good one. The situation is one open to opportunistic abuses on both sides. The innovator has incentives to overstate the value of the innovation, while the supplier has incentives to run with the technology should the innovation be a success.

Instances of both parties making irreversible capital commitments nevertheless exist. Apple's Laserwriter—a high-resolution laser printer that allows PC users to produce near-typeset-quality text and art-department graphics—is a case in point. Apple persuaded Canon to participate in the development of the Laserwriter by providing subsystems from its copiers—but only after Apple contracted to pay for a certain number of copier engines and cases. In short, Apple accepted a good deal of the financial risk in order to induce Canon to assist in the development and production of the Laserwriter. The arrangement appears to have been prudent, yet there were clearly hazards for both sides. It is difficult to write, execute, and enforce complex development contracts, particularly when the design of the new product is still floating. Apple was exposed to the risk that its co-innovator Canon would fail to deliver, and Canon was exposed to the risk that the Apple design and marketing effort would not succeed. Still, Apple's alternatives may have been rather limited, inasmuch as it did not command the requisite technology to go it alone.

In short, the current euphoria over strategic partnering may be partially misplaced. The advantages are being stressed (for example, McKenna 1985) without a balanced presentation of costs and risks. Briefly, there is the risk that the partner will not perform according to the innovator's perception of what the contract requires; there is the added danger that the partner may imitate the innovator's technology and attempt to compete with the innovator. This latter possibility is particularly acute if the provider of the complementary asset is uniquely situated with respect to the complementary asset in question and has the capacity to imitate the technology that the innovator is unable to protect. The innovator will then find that it has created a competitor who is better positioned than the innovator to take advantage of the market opportunity at hand. *Business Week* (1986: 57–59) has expressed concerns along these lines in its discussion of the "hollow corporation."[5]

It is important to bear in mind, however, that contractual or partnering strategies in certain cases are ideal. If the innovator's technology is well protected, and if what the partner has to provide is a generic capacity available from many potential partners, then the innovator will be able to maintain the upper hand while avoiding the costs of duplicating downstream capacity. Even if the partner fails to perform, adequate alternatives exist (by assumption, the partner's capacities are commonly available) so that the innovator's efforts to commercialize its technology successfully ought to proceed profitably.

Integration Modes

Integration, which by definition involves ownership, is distinguished from pure contractual modes in that it typically facilitates incentive alignment and control. If an innovator owns rather than rents the complementary assets needed to commercialize, then it is in a position to capture spillover benefits stemming from increased demand for the complementary assets caused by the innovation.

Indeed, an innovator might be in the position, at least before its innovation is announced, to buy up capacity in the complementary assets, possibly to its great subsequent advantage. If futures markets exist, simply taking forward positions in the complementary assets might suffice to capture much of the spillovers.

Even after the innovation is announced, the innovator may still be able to build or buy complementary capacities at competitive prices if the innovation has ironclad legal protection (that is, if the innovation is in a tight appropriability regime). However, if the innovation is not tightly protected and once out is easy to imitate, then securing control of complementary capacities is likely to be the key success factor, particularly if those capacities are in fixed supply—so-called bottlenecks. Distribution and specialized manufacturing competences often become bottlenecks.

As a practical matter, however, an innovator may not have the time to acquire or build the complementary assets that ideally it would like to control. This is particularly true when imitation is so easy that timing becomes critical. Additionally, the innovator may simply not have the financial resources to proceed. The implications of timing and cash constraints are summarized in Figure 9–9.

Accordingly, in weak appropriability regimes, innovators need to rank complementary assets as to their importance. If the complementary assets are critical, ownership is warranted although if the firm is cash constrained, a minority position may well represent a sensible tradeoff.

Needless to say, when imitation is easy, strategic moves to build or buy complementary assets that are specialized must occur with due reference to the moves of competitors. There is no point in moving to build a specialized asset, for instance, if one's imitators can do it more quickly and cheaply.

It is self-evident that if the innovator is already a large enterprise with many of the relevant complementary assets under its control, integration is not likely to be the issue that it might otherwise be, because the innovating firm will already control many of the relevant specialized and cospecialized assets. However, in industries experiencing rapid technological change, technologies advance so rapidly that it is unlikely that a single company will have the full range of expertise needed to bring advanced products to market in a timely and cost-effective fashion. Hence, the integration issue is not just a small-firm issue.

Integration versus Contract Strategies: An Analytic Summary

Figure 9–10 summarizes some of the relevant considerations in the form of a decision flow chart. It indicates that a profit-seeking innovator, confronted by weak intellectual property protection and the need to

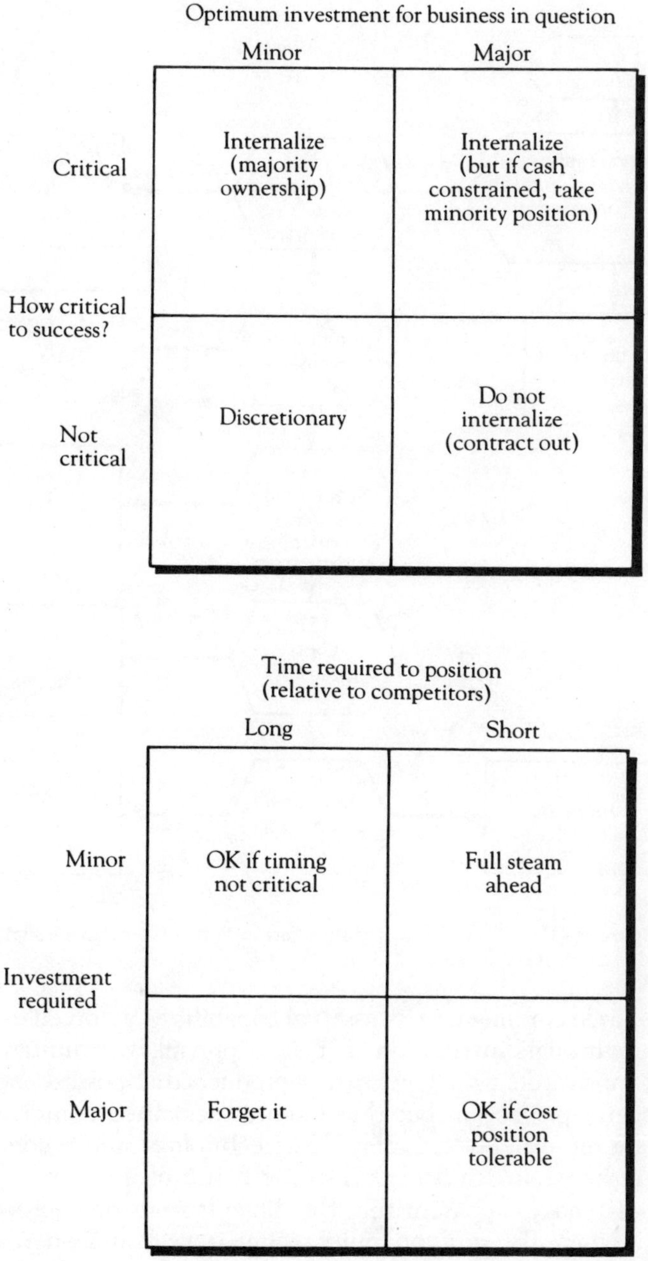

Optimum investment for business in question

	Minor	Major
Critical	Internalize (majority ownership)	Internalize (but if cash constrained, take minority position)
Not critical	Discretionary	Do not internalize (contract out)

How critical to success?

Time required to position (relative to competitors)

	Long	Short
Minor	OK if timing not critical	Full steam ahead
Major	Forget it	OK if cost position tolerable

Investment required

Figure 9–9. Specialized Complementary Assets and Weak Appropriability: Integration Calculus.

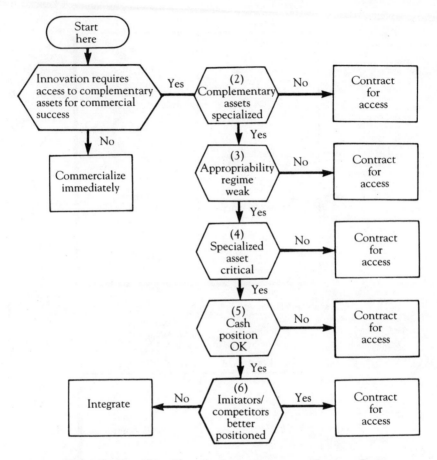

Figure 9–10.   Flow Chart for Integration versus Contract Design.

access specialized complementary assets or capabilities, is forced to expand its activities through integration if it is to prevail over imitators. Put differently, innovators who develop new products that possess poor intellectual property protection but that require specialized complementary capacities are more likely to parlay their technology into a commercial advantage rather than see it prevail in the hands of imitators.

Figure 9–10 makes apparent that the difficult strategic decisions arise in situations where the appropriability regime is weak and where specialized assets are critical to profitable commercialization. These situations, which in reality are very common, require that a fine-grained competitor analysis be part of the innovator's strategic assessment of its opportunities

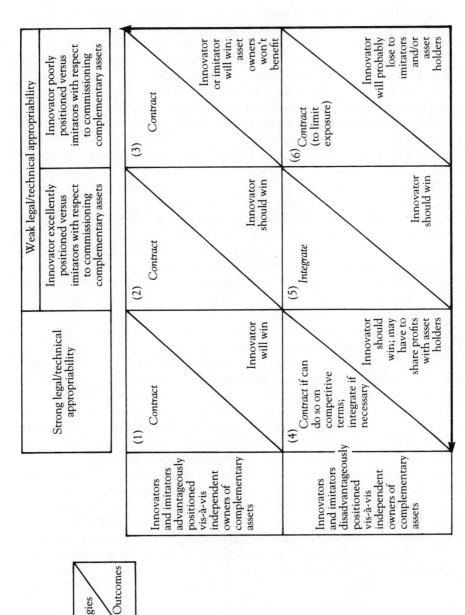

Figure 9–11. Contract and Integration Strategies and Outcomes for Innovators: Specialized Asset Case.

and threats. This is carried a step further in Figure 9–11, which looks only at situations where commercialization requires certain specialized capabilities. It indicates the appropriate strategies for the innovators and predicts the outcomes to be expected for the various players.

Three classes of players are of interest: innovators, imitators, and the owners of cospecialized assets (such as distributors). All three can potentially benefit or lose from the innovation process. The latter can potentially benefit from the additional business that the innovation may direct in the asset owner's direction. Should the asset turn out to be a bottleneck with respect to commercializing the innovation, the owner of the bottleneck facilities is obviously in a position to extract profits from the innovator or imitators.

The vertical axis in Figure 9–11 measures how those who possess the technology (the innovator or possibly its imitators) are positioned *vis-à-vis* those firms that possess required specialized assets. The horizontal axis measures the tightness of the appropriability regime, tight regimes being evidenced by ironclad legal protection coupled with technology that is simply difficult to copy; weak regimes offer little in the way of legal protection and the essence of the technology, once released, is transparent to the imitator. Weak regimes are further subdivided according to how the innovator and imitators are positioned *vis-à-vis* each other. This is likely to be a function of factors such as lead time and prior positioning in the requisite complementary assets.

Figure 9–11 makes it apparent that even when firms pursue the optimal strategy, other industry participants may take the jackpot. This possibility is unlikely when the intellectual property in question is tightly protected. The only serious threat to the innovator is where a specialized complementary asset is completely locked up, a possibility recognized in cell 4. This can rarely be done without the cooperation of government. But it frequently occurs, as when a foreign government closes off access to a foreign market, forcing the innovators to license to foreign firms with the effect of government cartelizing of the potential licensees. With weak intellectual-property protection, however, it is quite clear that the innovator will often lose out to imitators or asset holders even when the innovator is pursuing the appropriate strategy (cell 6). Clearly, incorrect strategies can compound problems. For instance, if innovators integrate when they should contract, a heavy commitment of resources will be incurred for little if any strategic benefit, thereby exposing the innovator to even greater losses than would otherwise be the case. On the other hand, if an innovator tries to contract for the supply of a critical capability when it should build the capability itself, it may well find it has nurtured an imitator better able to serve the market than the innovator itself.

Mixed Modes

The real world rarely provides extreme or pure cases. Decisions to integrate or license involve tradeoffs, compromises, and mixed approaches. It is not surprising, therefore, that the real world is characterized by mixed modes of organization involving judicious blends of integration and contracting. Sometimes mixed modes represent transitional phases. For instance, because of the convergence of computer and telecommunication technology, firms in each industry are discovering that they often lack the requisite technical capabilities in the other. Because the technological interdependence of the two requires collaboration among those who design different parts of the system, intense cross-boundary coordination and information flows are required. When separate enterprises are involved, agreement must be reached on complex protocol issues among parties who see their interests differently. Contractual difficulties can be anticipated because the selection of common technical protocols among the parties will often be followed by transaction-specific investments in hardware and software. There is little doubt that this was the motivation behind IBM's purchase of 15 percent of PBX manufacturer Rolm in 1983, a position that was expanded to 100 percent in 1984. IBM's stake in Intel, which began with a 12 percent purchase in 1982, is probably not a transitional phase leading to 100 percent purchase because both companies realize that the two corporate cultures are not very compatible, and IBM may not be so impressed with Intel's technology as it once was.

The CAT Scanner, the IBM PC, and NutraSweet: Insights from the Framework

EMI's failure to reap significant returns from the CAT scanner can be explained in large measure by reference to the concepts developed above. The scanner that EMI developed was of a technical sophistication much higher than would normally be found in a hospital, requiring a high level of training support and servicing. EMI had none of these capabilities, could not easily contract for them, and was slow to realize their importance. It most probably could have formed a partnership with a company like Siemens to access the requisite capabilities. Its failure to do so was a strategic error compounded by the very limited intellectual property protection that the law afforded the scanner. Although subsequent court decisions have upheld some of EMI's patent claims, once the product was in the market it could be reverse-engineered and its essential features copied. Two competitors, GE and Technicare, already possessed the complementary capabilities that the scanner required, and they were also technologically capable. In addition, both were experienced marketers of medical equipment and had reputations for quality, reliability, and service. GE and Technicare were thus able to commit their R&D resources to developing a competitive scanner, borrowing ideas from EMI's scanner, which they undoubtedly had access to through cooperative hospitals, and improving on it where they could while they rushed to market. GE began taking orders in 1976 and soon after made inroads on EMI. In 1977, concern for rising health-care costs caused the Carter administration to introduce certificate-of-need regulation, which required HEW's approval of expenditures on big-ticket items like CAT scanners. This severely cut the size of the available market.

By 1978 EMI had lost market-share leadership to Technicare, which was in turn quickly overtaken by GE. In October 1979 Godfrey Hounsfield of EMI shared the Nobel prize for invention of the CAT scanner. Despite this honor and the public recognition of its role in bringing this medical breakthrough to the world, the collapse of its scanner business forced EMI in the same year into the arms of a rescuer, Thorn Electrical Industries, Ltd. GE subsequently acquired what was EMI's scanner business from Thorn for what amounted to a pittance ("GE Gobbles a Rival in CAT Scanners" 1980). Though royalties continued to flow to EMI, the company had failed to capture the lion's share of the profits generated by the innovation it had pioneered and successfully commercialized.

If EMI illustrates how a company with outstanding technology and an excellent product can fail to profit from innovation while the imitators succeed, the story of the IBM PC indicates how a new product representing a very modest technological advance can yield remarkable returns to the developer.

The IBM PC, introduced in 1981, was a success despite the fact that the architecture was ordinary and the components standard. Philip Estridge's design team in Boca Raton, Florida, decided to use existing technology to produce a solid, reliable micro rather than state of the art. With a one-year mandate to develop a PC, Estridge's team could do little else.

However, the IBM PC did use what at the time was a new 16-bit microprocessor (the Intel 8088) and a new disk operating system (DOS) adapted for IBM by Microsoft. Other than the microprocessor and the operating system, the IBM PC incorporated existing micro standards and used off-the-shelf parts from outside vendors. IBM did write its own BIOS (Basic Input/Output System), which is embedded in ROM, but this was a relatively straightforward programming exercise.

The key to the PC's success was not the technology. It was the set of complementary assets that IBM either had or quickly assembled around the PC. In order to expand the market for PCs, there was a clear need for an expandable, flexible microcomputer system with extensive applications software. IBM could have based its PC system on its own patented hardware and copyrighted software. Such an approach would cause complementary products to be cospecialized, forcing IBM to develop peripherals and a comprehensive library of software in a very short time. Instead, IBM adopted what might be called an induced contractual approach. By adopting an open-system architecture, as Apple had done, and by making the operating-system information publicly available, a spectacular output of third-part software was induced. IBM estimated that by mid-1983, at least 3,000 hardware and software products were available for the PC (Gens and Christiansen 1983: 88). Put differently, IBM pulled together the complementary assets, particularly software, which success required, without even using contracts, let alone integration. This was despite the fact that the software developers were creating assets that in part cospecialized the IBM PC, at least in the first instance.

A number of special factors made this seem a reasonable risk to the software writers. A critical one was IBM's name and commitment to the project. The reputation behind the letters *I, B, M* is perhaps the greatest cospecialized asset the company possesses. The name implied that the product would be marketed and serviced in the IBM tradition. It guaranteed that PC-DOS would become an industry standard, so that the software business would not be dependent solely on IBM because emulators were sure to enter. It guaranteed access to retail distribution outlets on competitive terms. The consequence was that IBM was able to take a product that represented at best a modest technological accomplishment and turn it into a fabulous commercial success. The case demonstrates the role that complementary assets play in determining outcomes.

The spectacular success and profitability of G.D. Searle's NutraSweet is an uncommon story that is also consistent with the above framework. In 1982 Searle reported combined sales of $74 million for NutraSweet and its tabletop version, Equal. In 1983 this surged to $336 million. In 1985 NutraSweet sales exceeded $700 million (*Monsanto Annual Report* 1985) and Equal had captured 50 percent of the U.S. sugar-substitute market and was number one in five other countries.

NutraSweet, which is Searle's trade name for aspartame, has achieved rapid acceptance in each of its FDA-approved categories because of its good taste and ability to substitute directly for sugar in many applications. However, Searle's earnings from NutraSweet and the absence of a strategic challenge can be traced in part to Searle's clever strategy.

It appears that Searle has managed to establish an exceptionally tight appropriability regime around NutraSweet—one that may well continue for some time after the patent has expired. No competitor appears successfully to have invented around the Searle patent and commercialized an alternative, no doubt in part because the FDA approval process would have to begin anew for an imitator who was not violating Searle's patents. A competitor who tried to replicate the aspartame molecule with minor modification to circumvent the patent would probably be forced to replicate the hundreds of tests and experiments that proved aspartame's safety. Without patent protection, FDA approval would provide no shield against imitators that would come to market with an identical chemical and that could establish to the FDA that it was the same compound that had already been approved. Without FDA approval, on the other hand, the patent protection would be worthless, for the product would not be sold for human consumption.

Searle has aggressively pushed to strengthen its patent protection. The company was granted U.S. patent protection in 1970. It has also obtained patent protection in Japan, Canada, Australia, United Kingdom, France, West Germany, and a number of other countries. However, most of these patents carry a seventeen-year life. Because the product was approved for human consumption only in 1982, the seventeen-year patent life was effectively reduced to five. Recognizing the obvious importance of its patent, Searle pressed for and obtained special legislation in November 1984 extending the patent protection on aspartame for another five years. The United Kingdom provided a similar extension. In almost every other nation, however, 1987 will mark the expiration of the patent.

When the patent expires, however, Searle will still have several valuable assets to help keep imitators at bay. Searle has gone to great lengths to create and promulgate the use of its NutraSweet name and a distinctive *Swirl* logo on all goods licensed to use the ingredient. The company has also developed the *Equal* trade name for a tabletop version of the sweetener. Trademark law in the United States provides protection against unfair competition in branded products as long as the owner of the mark continues to use it. Both the NutraSweet and Equal trademarks will become essential assets when the patents on aspartame expire. Searle may well have convinced consumers that the only real form of sweetener is NutraSweet/Equal. Consumers know most other artificial sweeteners by their generic names—saccharin and cyclamates.

Clearly, Searle is trying to build a position in complementary assets to prepare for the competition that will surely arise. Searle's joint venture with Ajinomoto ensures them access to that company's many years of experience in the production of biochemical agents. Much of this knowledge is associated with techniques for distillation and synthesization of the delicate hydrocarbon compounds that are the ingredients in NutraSweet and is therefore more tacit than codified. Searle has begun to put these techniques to use in its own $160 million Georgia production facility. It can be expected that Searle will use trade secrets to the maximum to keep this know-how proprietary.

By the time its patent expires, Searle's extensive research into production techniques for L-phenylalanine and its eight years of experience in the Georgia plant should give it significant cost advantage over potential aspartame competitors. Trade-secret protection, unlike patents, has no fixed lifetime and may well sustain Searle's position for years to come.

Searle, moreover, has wisely avoided renewing contracts with suppliers when they have expired. Had Searle subcontracted manufacturing for NutraSweet, it would have created a manufacturer who would have been in a position to enter the aspartame market itself or to team up with a marketer of artificial sweeteners. By keeping manufacturing in-house and by developing a valuable trade name, Searle has a good chance of protecting its market position from dramatic inroads once patents expire. Clearly, Searle seems to be astutely aware of the importance of maintaining a tight appropriability regime and using cospecialized assets strategically.

Implications for R&D Strategy, Industry Structure, and Trade Policy

Allocating R&D Resources

The analysis so far assumes that the firm has developed an innovation for which a market exists. It indicates the strategies that the firm must follow to maximize its share of industry profits relative to imitators and other competitors. There is no guarantee of success even if optimal strategies are followed.

The innovator can improve its total return to R&D, however, by adjusting its R&D investment portfolio to maximize the probability that technological discoveries will emerge that either are easy to protect with existing intellectual property law or require for commercialization cospecialized assets already within the firm's repertoire of capabilities. Put differently, if an innovating firm does not target its R&D resources toward new products and processes that it can commercialize advantageously relative to potential imitators or followers, it is unlikely to profit from its investment in R&D. In this sense, a firm's history—and the assets it already has in place—ought to condition its R&D investment decisions. Clearly, an innovating firm with considerable assets already in place is free to strike out in new directions as long as in doing so it is aware of the kinds of capabilities required to commercialize the innovation successfully. It is therefore rather clear that the R&D investment decision cannot be divorced from the strategic analysis of markets and industries and the firm's position within them.

Small Firm versus Large Firm Comparisons

Business commentators often remark that many small entrepreneurial firms that generate new, commercially valuable technology fail while large multinational firms, often with less meritorious records with respect to innovation, survive and prosper. One set of reasons for this phenomenon is now clear. Large firms are more likely to possess the relevant specialized and cospecialized assets within their boundaries at the time of new-product introduction. They can therefore do a better job of milking their technology, however meager, to maximum advantage. Small domestic firms are less likely to have the relevant specialized and cospecialized assets within their boundaries and so either will have to incur the expense of trying to build them or will have to try to develop coalitions with competitors/owners of the specialized assets.

Regimes of Appropriability and Industry Structure

In industries where legal methods of protection are effective or where new products are just hard to copy, the strategic necessity for innovating firms to integrate into cospecialized assets would appear to be less compelling than in industries where legal protection is weak. In cases where legal protection is weak or nonexistent, the control of cospecialized assets will be needed for long-run survival.

In this regard, it is instructive to examine the U.S. drug industry (Temin 1979). Beginning in the 1940s, the U.S. Patent Office began, for the first time, to grant patents on certain natural substances that involved difficult extraction procedures. Thus, in 1948 Merck received a patent on Streptomycin, which was a natural substance. However, it was not the extraction process but the drug itself that received the patent. Hence, patents were important to the drug industry in terms of what could be patented (drugs), but they did not prevent imitation (Temin 1979: 436). Sometimes just changing one molecule will enable a company to come up with a different substance that does not violate the patent. Had patents been more all inclusive—and I am not suggesting they should be—licensing would have been an effective mechanism for Merck to extract profits from its innovation. As it turns out, the emergence of close substitutes—coupled with FDA regulation that had the *de facto* effect of reducing the elasticity of demand for drugs—placed high rewards on a product differentiation strategy. This required extensive marketing, including a salesforce that could directly contact doctors, who were the purchasers of drugs through their ability to create prescriptions.[6] The result was exclusive production (that is, the earlier industry practice of licensing was dropped) and forward integration into marketing (the relevant cospecialized asset).

Generally, if legal protection of the innovator's profits is secure, innovating firms can select their boundaries based simply on their ability to identify user needs and respond to those through research and development. The weaker the legal methods of protection, the greater the incentive to integrate into the relevant cospecialized assets. Hence, as industries in which legal protection is weak begin to mature, integration into innovation-specific cospecialized assets will occur. Often this will take the form of backward, forward, and lateral integration. (Conglomerate integration is not part of this phenomenon.) For example, IBM's purchase of Rolm can be seen as a response to the impact of technological change on the identity of the cospecialized assets relevant to IBM's future growth.

Industry Maturity, New Entry, and History

As technologically progressive industries mature and a greater proportion of the relevant cospecialized assets are brought in under the corporate umbrellas of incumbents, new entry becomes increasingly difficult. Moreover, when it does occur, it is more likely to involve coalition formation very early on. Incumbents will for sure own the cospecialized assets, and new entrants will find it necessary to forge links with them. Here lies the explanation for the sudden surge in strategic partnering now occurring internationally, and particularly in the computer and telecommunications industry. Note that it should not be interpreted in anticompetitive terms. Given existing industry structure, coalitions ought to be seen not as attempts to stifle competition but as mechanisms for lowering entry requirements for innovators.

In industries where there has occurred technological change of a kind that required deployment of specialized or cospecialized assets at the time, a configuration of firm boundaries may well have arisen that no longer has compelling efficiencies. Considerations that once dictated integration may no longer hold, yet there may not be strong forces leading to divestiture. Hence, existing firm boundaries in some industries—especially those where the technological trajectory and attendant specialized asset requirements have changed—may be rather fragile. In short, history matters in terms of understanding the structure of the modern business enterprise. Existing firm boundaries cannot always be assumed to have obvious rationales in terms of today's requirements.

The Importance of Manufacturing to International Competitiveness

Practically all forms of technological know-how must be embedded in goods and services to yield value to the consumer. An important policy issue for the innovating nation is whether the identity of the firms and nations performing this function matter.

In a world of tight appropriability and zero transactions cost—the world of neoclassical trade theory—it is a matter of indifference whether an innovating firm has an in-house manufacturing capability, domestic or foreign. It can simply engage in arm's-length contracting (patent licensing, know-how licensing, co-production, and so forth) for the sale of the output of the activity in which it has a comparative advantage (in this case R&D) and will maximize returns by specializing in what it does best.

However, in a regime of weak appropriability, especially where the requisite manufacturing assets are specialized to the innovation, which is often the case, participation in manufacturing may be necessary if an innovator is to appropriate the rents from its innovation. Hence, if an innovator's manufacturing costs are higher than those of its imitators, the innovator may well end up ceding the lion's share of profits to the imitator.

In a weak appropriability regime, low-cost imitator-manufacturers may end up capturing all of the profits from innovation. In a weak appropriability regime where specialized manufacturing capabilities are required to produce new products, an innovator with a manufacturing disadvantage may find that its advantage at early-stage research and development will have no commercial value. This will eventually cripple the innovator, unless it is assisted by governmental processes. For example, it appears that one of the reasons why U.S. color television manufacturers did not capture the lion's share of the profits from the innovation, for which RCA was primarily responsible, was that RCA and its U.S. licensees were not competitive at manufacturing. In this context, concerns that the decline of manufacturing threatens the entire economy appear to be well founded.

A related implication is that as the technology gap closes, the basis of competition in an industry will shift to the cospecialized assets. This appears to be what is happening in microprocessors. Intel is no longer ahead technologically. As Gordon Moore, CEO of Intel points out ("Institutionalizing the Revolution" 1986: 35),

> Take the top 10 [semiconductor] companies in the world . . . and it is hard to tell at any time who is ahead of whom. . . . It is clear that we have to be pretty damn close to the Japanese from a manufacturing standpoint to compete.

It is not just that strength in one area is necessary to compensate for weakness in another. As technology becomes more public and less proprietary through easier imitation, strength in manufacturing and other capabilities is necessary to derive advantage from whatever technological advantages an innovator may possess.

Put differently, the notion that the United States can adopt a designer role in international commerce while letting independent firms in other countries such as Japan, Korea, Taiwan, or Mexico do the manufacturing is unlikely to be viable as a long-run strategy because profits will accrue primarily to the low-cost manufacturers (by providing a larger sales base over which they can exploit their special skills). Where imitation is easy, and even where it is not, there are obvious problems in transacting in the market for know-how—problems that are described in more detail elsewhere (Teece 1981). In particular, there are difficulties in pricing an intangible asset whose true performance features are difficult to ascertain *ex ante*.

The trend in international business towards what Miles and Snow (1986) call *dynamic networks*—characterized by vertical disintegration and contracting—thus ought to be viewed with concern. (*Business Week* 1986 has referred to the same phenomenon as the *hollow corporation*.) Dynamic networks may not so much reflect innovative organizational forms as the disassembly of the modern corporation because of deterioration in national capacities, in manufacturing particularly, that are complementary to technological innovation. Dynamic networks may therefore signal not so much the rejuvenation of U.S. enterprise as its piecemeal demise.

How Trade and Investment Barriers Can Affect Innovators' Profits

In regimes of weak appropriability, governments can move to shift the distribution of the gains from innovation away from foreign innovators and toward domestic firms by denying innovators ownership of specialized assets. The foreign firm, which by assumption is an innovator, will be left with the option of selling its intangible assets in the market for know-how if both trade and investment are foreclosed by government policy. This option may appear better than the alternative (no remuneration at all from the market in question). Licensing may then appear profitable but only because access to the complementary assets is blocked by government.

Thus when an innovating firm generating profits needs to access complementary assets abroad, host governments, by limiting access, can sometimes milk the innovators for a share of the profits, particularly that portion that originates from sales in the host country. However, the ability of host governments to do so depends importantly on how critical the host country's assets are to the innovator. If the cost and infrastructure characteristics of the host country are such that it is the world's lowest-cost manufacturing site and if domestic industry is competitive, then by acting as a *de facto* monopsonist, the host-country government ought to be able to adjust the terms of access to the complementary assets so as to appropriate a greater share of the profits generated by the innovation.[7] If, on the other hand, the host country offers no unique complementary assets, except access to its own market, restrictive practices by the government will only redistribute profits with respect to domestic rather than worldwide sales.

Implications for the International Distribution of the Benefits from Innovation

The above analysis indicates that innovators who do not have access to the relevant specialized and cospecialized assets may end up ceding profits to imitators and other competitors or simply to the owners of the specialized or cospecialized assets. Even when the specialized assets are possessed by the innovating firm, they may be located abroad. Under these circumstances, foreign factors of production are likely to benefit from the research and development activities. There is little doubt, for instance, that the inability of many U.S. multinationals to sustain competitive manufacturing in the United States is resulting in declining returns to U.S. labor. Stockholders and top management probably do as well, if not better, when a multinational accesses cospecialized assets in the firm's foreign subsidiaries; however, if there is unemployment in the factors of production supporting the specialized and cospecialized assets in question, then the foreign factors of production will benefit from innovation originating beyond national borders. This speaks to the importance to innovating nations of maintaining competence and competitiveness in the assets that complement technological innovation, manufacturing being a case in point. It also speaks to the importance to innovating nations of enhancing the protection afforded worldwide to intellectual property.

However, it must be recognized that there are inherent limits to the legal protection of intellectual property and that business and national strategy are therefore likely to be critical factors in determining how the gains from innovation are shared worldwide. By making the correct strategic decision, innovating firms can move to protect the interests of stockholders; however, to ensure that domestic rather than foreign cospecialized assets capture the lion's share of the externalities spilling over to complementary assets, the supporting infrastructure for those complementary assets must not be allowed to decay. In short, if a nation has prowess at innovation, then in the absence of ironclad protection for intellectual property, it must maintain well-developed complementary assets if it is to capture the spillover benefits from innovation.

Conclusion

This analysis has attempted to synthesize from recent research in industrial organization and strategic management a framework within which to analyze the distribution of the profits from innovation. The framework indicates that the boundaries of the firm are an important strategic variable for innovating firms. The ownership of complementary assets, particularly when they are specialized or cospecialized, helps establish who wins and who loses from innovation. Imitators can often outperform innovators if they are better positioned with respect to critical complementary assets. Hence, public policy aimed at promoting innovation must focus not only on R&D but also on complementary assets as well as the underlying infrastructure. If government decides to stimulate innovation, it would seem important to clear away barriers that would impede the development of complementary assets that tend to be specialized or cospecialized to innovation. To fail to do so would cause an unnecessarily large portion of the profits from innovation to flow to imitators and other competitors. When these firms lie beyond national borders, there are obvious implications for the international distribution of income.

When applied to world markets, results similar to those obtained from the new trade theory are suggested by the framework. In particular, tariffs and other restrictions on trade can in some cases injure innovating firms while simultaneously benefiting protected firms when they are imitators. However, the propositions suggested by the framework are particularized to appropriability regimes, suggesting that economywide conclusions will be illusive. The policy conclusions derivable for commodity petrochemicals, for instance, are likely to be different from those that would be arrived at for semiconductors.

The approach also suggests that the product life cycle model of international trade will play itself out very differently in different industries and markets, in part according to appropriability regimes and the nature of the assets that need to be employed to convert a technological success into a commercial one. Whatever its limitations, the approach establishes that it is not so much the structure of markets but the structure of firms, particularly the scope of their boundaries, coupled with national policies with respect to the development of complementary assets, that determines the distribution of the profits among innovators and imitator/followers.

NOTES

1. The EMI story is summarized in Martin (1984).
2. Robert D. Kennedy, executive vice president of Union Carbide, quoted in *Chemical Week.*
3. Comment attributed to Peter Olson III, IBM's director of business development, as reported in "The Strategy behind IBM's Strategic Alliances" (1985: 126).
4. Comment attributed to Norman Farquhar, Cipher's vice president for strategic development, as reported in *Electronic Business* (1985: 128).
5. *Business Week* uses the term to describe a corporation that lacks in-house manufacturing capability.
6. In the period before FDA regulation, all drugs other than narcotics were available over the counter. Because the end user could purchase drugs directly, sales were price sensitive. Once prescriptions were required, this price sensitivity collapsed; the doctors not only did not have to pay for the drugs, but in most cases they were unaware of the prices of the drugs that they were prescribing.
7. If the host-country market structure is monopolistic in the first instance, private actors might be able to achieve the same benefit. What government can do is to force collusion of domestic enterprises to their mutual benefit.

REFERENCES

"A Bad Aftertaste." 1985. *Business Week.* (July 15).
Abernathy, W.J., and J.M. Utterback. 1978. "Patterns of Industrial Innovation." *Technology Review* 80(7) (January/July): 40–47.
Clark, Kim B., 1985. "The Interaction of Design Hierarchies and Market Concepts in Technological Evolution." *Research Policy* 14: 235–251.
Dosi, G. 1982. "Technological Paradigms and Technological Trajectories." *Research Policy* 11(3): 147–62.
"GE Gobbles a Rival in CT Scanners." 1980. *Business Week.* (May 19): p. 47.
Gens, F., and C. Christiansen. 1983. "Could 1,000,000 IBM PC Users Be Wrong?" *Byte* (November): 88.
"The Hollow Corporation." 1986. *Business Week* (March 3): 57–59.
"Institutionalizing the Revolution." 1986. *Forbes* (June 16): 35.
Kuhn, Thomas. 1970. *The Structure of Scientific Revolutions*, 2d ed. Chicago: University of Chicago Press.
Levin, R., A. Klevorick, N. Nelson, and S. Winter. 1984. "Survey Research on R&D Appropriability and Technological Opportunity." Unpublished manuscript, Yale University.
Martin, Michael. 1984. *Managing Technical Innovation and Entrepreneurship.* Reston, Va.: Reston.
McKenna, Regis. 1985. "Market Positioning in High Technology." *California Management Review* 27(3) (Spring): 82–108.
Miles, R.E., and C.C. Snow. 1986. "Network Organizations: New Concepts for New Forms." *California Management Review* 28(3) (Spring): 62–73.
Monsanto Annual Report. 1985. St. Louis, Mo.
Norman, David A. 1986. "Impact of Entrepreneurship and Innovations on the Distribution of Personal Computers." In *The Positive Sum Strategy*, edited by R. Landau and N. Rosenberg, pp. 437–39. Washington, D.C.: National Academy Press.
"The Strategy behind IBM's Strategic Alliances." 1985. *Electronic Business* 11(19) (October 1): 126–29.

Teece, D.J. 1981. "The Market for Know-How and the Efficient International Transfer of Technology." *Annals of the American Academy of Political and Social Science* 458 (November): 81–96.

Temin, P. 1979. "Technology, Regulation, and Market Structure in the Modern Pharmaceutical Industry." *Bell Journal of Economics* 10(2) (Autumn): 429–46.

Williamson, O.E. 1975. *Markets and Hierarchies.* New York: Free Press.

10 SUBSTITUTES FOR STRATEGY

Karl E. Weick

A little strategy goes a long way. Too much can paralyze or splinter an organization. That conclusion derives from the possibility that strategy-like outcomes originate from sources other than strategy. Adding explicit strategy to these other tacit sources of strategy can be self-defeating and reduce effectiveness (Bresser and Bishop 1983). Thus, the focus of this chapter is substitutes for strategy.

The model for this exercise is the concept in the leadership literature of substitutes for leadership (Kerr and Jermier 1978). Substitutes are conditions that either neutralize what leaders do or perform many of the same functions they would. Substitutes include characteristics of subordinates (ability, knowledge, experience, training, professional orientation, indifference toward organizational rewards), characteristics of the task (unambiguous, routine, provides its own feedback, intrinsically satisfying), and characteristics of the organization (high formalization, highly specified staff functions, closely knit cohesive groups, organizational rewards not controlled by leaders, spatial distance between subordinates and superiors). Leadership has less impact when one or more of these conditions obtains. It is not that the situation is devoid of leadership; rather, the leadership is done by something else.

It seems reasonable to work analogically and investigate the extent to which it is possible to create substitutes for strategies.

If pressed to define *strategy*, I am tempted to adopt DeBono's (1984: 143) statement that "strategy is good luck rationalized in hindsight," but I am also comfortable with a definition much like Robert Burgelmann's (1983)—namely, "strategy is a theory about the reasons for past and current success of the firm." Both of my definitional preferences differ sharply from Chandler's (1962) classic definition of *strategy*—"The determination of the basic long-term goals and objectives of an enterprise, and the adoption of courses of action and the allocation of resources necessary for carrying out these goals."

Definitions notwithstanding, I can best show what I think strategy is by describing an incident that happened during military maneuvers in Switzerland. The young lieutenant of a small Hungarian detachment in the Alps sent a reconnaissance unit into the icy wilderness. It began to snow immediately, snowed for two days, and the unit did not return. The lieutenant suffered, fearing that he had dispatched his own people to death. But the third day the unit came back. Where had they been? How had they made their way? Yes, they said, we considered ourselves lost and waited for the end. And then one of us found a map in his pocket. That calmed us down. We pitched camp, lasted out the snowstorm, and then with the map we discovered our bearings. And here we are. The lieutenant borrowed this remarkable map and had a good look at it. He discovered to his astonishment that it was not a map of the Alps, but a map of the Pyrenees.

This incident raises the intriguing possibility that when you are lost, any old map will do. Extended to the issue of strategy, maybe when you are confused, any old strategic plan will do.

Strategic plans are a lot like maps. They animate people and they orient people. Once people begin to act, they generate tangible outcomes in some context, and this helps them discover what is occurring, what needs to be explained, and what should be done next. Managers keep forgetting that it is what they do, not what they plan, that explains their success. They keep giving credit to the wrong thing—namely, the plan— and having made this error, they then spend more time planning and less time acting. They are astonished when more planning improves nothing.

Kirk Downey has suggested that the Alps example is a success story for two quite specific reasons. First, the troops found a specific map that was relevant to their problem. Had they found a map of Disneyland rather than a map of the Pyrenees their problem would have deepened materially. Second, the troops had a purpose—that is, they wanted to go back to their base camp—and it was in the context of this purpose that the map took on meaning as a means to get them back. These conditions, however, do not negate the basic theme that meaning lies in the path of the action. A map of Disneyland makes it harder to develop a shared understanding of what has happened and where we have been, but if it does not inhibit action and observation, some clearer sense of the situation may emerge as action proceeds.

When I described the incident of using a map of the Pyrenees to find a way out of the Alps to Bob Engel, the executive vice president and treasurer of Morgan Guaranty, he said, "Now, that story would have been really neat if the leader out with the lost troops had known it was the wrong map and still been able to lead them back."

What is interesting about Engel's twist to the story is that he has described the basic situation that most leaders face. Followers are often lost and even the leader is not sure where to go. All the leader knows is that the plan or the map he has in front of him is not sufficient by itself to get them out. What he has to do, when faced with this situation, is instill some confidence in people, get them moving in some general direction, and be sure they look closely at what actually happens, so that they learn where they were and get some better idea of where they are and where they want to be.

If you get people moving, thinking clearly, and watching closely, events often become more meaningful. For one thing, a map of the Pyrenees can still be a plausible map of the Alps because in a very general sense, if you have seen one mountain range, you have seen them all (readers can test this assertion for themselves by examining "A Traveler's Map of the Alps" in the April 1985 issue of *National Geographic Magazine*). The Pyrenees share some features with the Alps, and if people pay attention to these common features, they may find their way out. For example, most mountain ranges are wet on one side and dry on the other. Water flows down rather than up. There is a prevailing wind. There are peaks and valleys. There is a highest point, and then the peaks get lower and lower until there are foothills.

Just as it is true that if you have seen one mountain range you have seen them all, it also is true that if you have seen one organization you have seen them all. Any old plan will work in an organization because people usually learn by trial and error, some people listen and some people talk, people want to get somewhere and have some general sense of where they now are, 20 percent of the people will do 80 percent of the work (and vice versa), and if you do something for somebody, they are more likely to do something for you. Given these general features of most organizations, any old plan is often sufficient to get this whole mechanism moving, which then makes it possible to learn what is going on and what needs to be done next.

The generic process involved is that meaning is produced because the leader treats a vague map or plan as if it had some meaning, even though he knows full well that the real meaning will come only when people respond to the map and do something. The secret of leading with a bad map is to create a self-fulfilling prophecy. Having predicted that the group will find its way out, the leader creates the combination of optimism and action that allows people to turn their confusion into meaning and find their way home.

There are plenty of examples in industry where vague plan and projects provide an excuse for people to act, learn, and create meaning.

The founders of Banana Republic, the successful mail order clothier, started their business by acting in an improbable manner. They bought uniforms from overthrown armies in South America and advertised these items in a catalog, using drawings rather than photographs. All of these actions were labeled poor strategy by other mail order firms. When these three actions were set in motion, however, they generated responses that no one expected (because no one had tested them) and created a belated strategy as well as a distinct niche for Banana Republic.

Tuesday Morning, an off-price retailing chain that sells household and gift items, opens its stores when they have enough merchandise to sell and then closes them until they get the next batch. As managers followed this pattern, they discovered that customers love grand openings and that anticipation would build between closings over when the store would open again and what it would contain. These anticipations were sufficiently energizing that stores that opened intermittently for four to eight weeks sold more than equivalent stores that were open year round.

The Microelectronics and Computer Technology Corporation (MCC) consortium in Austin, Texas, is a clear example of the sequence in which vague projects trigger sufficient action that vagueness gets removed. A key Texas state official described MCC as "an event, not a company." Bidding for MCC to locate in Texas became a vehicle to pull competing Texas cities together. It also became a vehicle to tell out-of-state people, "We are a national and an international force, not just a regional force, and not just a land of cowboys and rednecks." MCC became a tangible indication that Texas was growing, maturing, and on its way up. MCC's criteria for a good site became defining characteristics of what Austin was as a city, though Austinites did not realize they had this identity before. MCC said in its specifications that it did not want to locate where everyone thinks they know how high-tech R&D should be done. Texas thus "discovered" that its backwardness was in fact one of its biggest assets.

Acquiring MCC became a strategy to strengthen Texas, but only quite late, when more and more problems were seen to be solved if it landed in Austin. The action of bidding for MCC fanned out in ways that people had not anticipated. The point is, if action is decoupled from strategy, then people have a better chance to be opportunistic, to discover missions and resources they had no idea existed.

So far three themes have been introduced: (1) that action clarifies meaning; (2) that the pretext for the action is of secondary importance;

(3) and that strategic planning is only one of many pretexts for meaning-generation in organizations. To clarify some ways in which action can substitute for strategy, we will look more closely at the dynamics of confidence and improvisation.

Confidence as Strategy

In managerial work, thought precedes action, but the kind of thought that often occurs is not detailed analytical thought addressed to imagined scenarios in which actions are tried and options chosen. Instead, thought precedes action in the form of much more general expectations about the orderliness of what will occur.

Order is present, not because extended prior analysis revealed it but because the manager anticipates sufficient order that she wades into the situation, imposes order among events, and then "discovers" what she had imposed. The manager "knew" all along that the situation would make sense. This was treated as a given. Having presumed that it would be sensible, the manager than acts confidently and implants the order that was anticipated.

Most managerial situations contain gaps, discontinuities, loose ties among people and events, indeterminacies, and uncertainties. These are the gaps that managers have to bridge. It is the contention of this argument that managers first think their way across these gaps and then, having tied the elements together cognitively, actually tie them together when they act and impose covariation. This sequence is similar to sequences associated with self-fulfilling prophecies (see Snyder, Tanke, and Berscheid 1977).

Thus presumptions of logic are forms of thought that are crucial for their evocative qualities. The presumption leads people to act more forcefully, the more certain the presumption. Strong presumptions (such as, "I know that these are the Pyrenees") lead to strong actions that impose considerable order. Weaker presumptions lead to more hesitant actions, which means either that the person will be more influenced by the circumstances that are already present or that only weak order will be created.

Presumptions of logic are evident in the chronic optimism often associated with managerial activity. This optimism is conspicuous in the case of companies that are in trouble, but it is also evident in more run-of-the-mill managing. Optimism is one manifestation of the belief that situations will have made sense. William James (1956) described the faith that life is worth living that generates the action that then makes life worth living. Optimism is not necessarily a denial of reality. Instead it may be the belief that makes reality possible.

Presumptions of logic should be prominent among managers because of the climate of rationality in organizations (Staw 1980). Presumptions should be especially prominent when beliefs about cause and effect linkages are unclear (Thompson 1964: 336). Thompson labels the kind of managing that occurs when there are unclear preferences and unclear cause/effect beliefs *inspiration*. It is precisely in the face of massive uncertainty that beliefs of some sort are necessary to evoke some action, which can then begin to consolidate the situations. To inspire is to affirm realities, which then are more likely to materialize if they are sought vigorously. That sequence may be the essence of managing.

Examples of the effect of presumptions are plentiful. A male who believes he is telephoning an attractive female speaks more warmly, which evokes a warm response from her, which confirms the original stereotype that attractive women are sociable (Snyder, Tanke, and Berscheid 1977). A new administrator, suspecting that old-timers are traditional, seeks ideas from other sources, which increases the suspicion of old-timers and confirms the administrator's original presumption (Warwick, 1975). People who presume that no one likes them approach a new gathering in a stiff, distrustful manner, which evokes the unsympathetic behavior they presumed would be there (Watzlawick, Beavin, and Jackson 1967: 98–99). A musician who doubts the competence of a composer plays his music lethargically and produces the ugly sound that confirms the original suspicion (Weick, Gilfillan, and Keith 1973).

In each case, an initial presumption (she is sociable, they are uncreative, people are hostile, he is incompetent) leads people to act forcibly (talk warmly, seek ideas elsewhere, behave defensively, ignore written music), which causes a situation to become more orderly (warmth is exchanged, ideas emerge, hostility is focused, music becomes simplistic), which then makes the situation easier to interpret, thereby confirming the original presumption that it will have been logical.

This sequence is common among managers because managerial actions are almost ideally suited to sustain self-fulfilling prophecies (Eden 1984). Managerial actions are primarily oral, face-to-face, symbolic, presumptive, brief, and spontaneous (McCall and Kaplan 1985). These actions have a deterministic effect on many organizational situations because those situations are less tightly coupled than are the confident actions directed at them. The situations are loosely coupled, subject to multiple interpretations, monitored regularly by only a handful of people, and deficient in structure.

Thus a situation of basic disorder becomes more orderly when people overlook the disorder and presume orderliness, then act on this presumption, and finally rearrange its elements into a more meaningful arrangement that confirms the original presumption. It is suggested that typical

managerial behavior is more likely to create rather than disrupt this sequence. Thus, a manager's preoccupation with rationality may be significant less for its power as an analytic problemsolving tool than for its power to induce action that eventually implants the rationality that was presumed when the sequence started.

The lesson of self-fulfilling prophecies for students of strategy is that strong beliefs that single out and intensify consistent action can bring events into existence (see Snyder 1984). Whether people are called fanatics, true believers, or the currently popular phrase *idea champions*, they all embody what looks like strategy in their persistent behavior. Their persistence carries the strategy; the persistence is the strategy. True believers impose their view on the world and fulfill their own prophecies. Note that this makes strategy more of a motivational problem than a cognitive forecasting problem.

An argument can be made that the so-called computer revolution is an ideal exhibit of confidence as strategy. The revolution is as much vendor-driven as it is need-driven. The revolution can be viewed as solutions in search of problems people never knew they had. Vendors had more forcefulness, confidence, and focus than did their customers, who had only a vague sense that things were not running right, although they could not say why. Vendors defined the unease as a clear problem in control and information distribution, a definition that was no worse than any other diagnosis that was available.

To say that it was IBM's strategy to be forceful is to miss the core of what actually happened. The key point is that IBM's strategy worked after it became self-confirming, when it put an environment in place. A common error is that the strategic plan is valued because it looks like it correctly forecast a pent-up demand for computers. Actually, it did no such thing. Instead, the plan served as a pretext for people to act forcefully and impose their view of the world. Once they imposed, enacted, and stabilized that view and once it was accepted, then more traditional procedures of strategic planning could be made to work because they were directed at more predictable problems in a more stable environment. What gets missed by strategy analysts is that proaction precedes reaction. Strategic planning works only after forceful action has hammered the environment into shape so that it is less variable and so that conventional planning tools can now be made to work. Because the constrained environment contains demands, opportunities, and problems that were imposed during proaction, proaction, not planning, predicts what the organization has to contend with.

To see how self-fulfilling prophecies can mimic strategy and affect the direction of behavior, consider the problem of regulation. Although companies groan about the weight of regulation, data (McCaffrey 1982) suggest that regulators do not have their act together and are loosely coupled relative to the tightly coupled organizations and lawyers they try to regulate. Thus, many organizations ironically create the regulators who control them. The way they do this is a microcosm of the point being made here about confidence.

If a firm treats regulators as if they are unified and have their act together, then the firm gets its own act together to cope with the focused demands that are anticipated from the regulators. As the firm gets its act together it becomes a clearer target that is easier for the regulators to monitor and control. Concerted action undertaken by the firm to meet anticipated action from regulators now makes it possible for regulators to do something they could not have done when the firms were more diffuse targets.

A confident definition of regulatory power, confidently imposed, stabilizes the regulation problem for a firm. The irony is that the faulty prophecy brings the problem into existence more sharply than it ever was before confident behavior was initiated. The firm has become easier to regulate by virtue of its efforts to prevent regulation.

Environments are more malleable than planners realize. Environments often crystallize around prophecies, presumptions, and actions that unfold while planners deliberate. Guidance by strategy often is secondary to guidance by prophecies. These prophecies are more likely to fulfill themselves when they are in the heads of fanatics who work in environments where the definition of what is occurring can be influenced by confident assertions.

Thus, presumptions can substitute for strategy. We assume co-workers know where they are going, they assume the same for us, and both of us presume that the directions in which we both are going are roughly similar. A presumption does not necessarily mean that whatever is presumed actually exists. We often assume that people agree with us without ever testing that assumption. Vague strategic plans help because we never have to confront the reality of our disagreements. And the fact that those disagreements persist undetected is not necessarily a problem because those very differences provide a repertoire of beliefs and skills that allow us to cope with changing environments. When environmental change is rapid, diverse skills and beliefs are the solution, not the problem.

Improvisation as Strategy

Much of my thinking about organizations (such as Weick 1979) uses the imagery of social evolution, but there is a consistent bias in the way I use that idea. I consistently argue that the likelihood of survival goes up when variation increases, when possibilities multiply, when trial and error become more diverse and less stylized, when people become less repetitious, and when creativity becomes supported. Notice that variation, trial and error, and doing things differently all imply that what you already know, including your strategic plan, is not sufficient to deal with present circumstances.

When it is assumed that survival depends on variation, then a strategic plan becomes a threat because it restricts experimentation and the chance to learn that old assumptions no longer work. Furthermore, I assume that whatever direction strategy gives can be achieved just as easily by improvisation.

Improvisation is a form of strategy that is misunderstood. When people use jazz or improvisational theater to illustrate improvising, they usually forget that jazz consists of variations on a theme and improvisational theater starts with a situation. Neither jazz nor improvisational theater are anarchic. Both contain some order, but it is underspecified.

To understand improvisation as strategy is to understand the order within it. And what we usually miss is the fact that a little order can go a long way.

For example, we keep underestimating the power of corporate culture because it seems improbable that something as small as a logo, a slogan, a preference (Geneen's one unshakable fact), a meeting agenda, or a Christmas party could have such a large effect. The reason these symbols are so powerful is that they give a general direction and a frame of reference that are sufficient. In the hands of bright, ambitious, confident people who have strong needs to control their destinies, general guidelines are sufficient to sustain and shape improvisation without reducing perceived control.

If improvisation is treated as a natural form of organizational life, then we become interested in a different form of strategy than we have seen before. This newer form I will call a *just-in-time strategy*. Just-in-time strategies are distinguished by less investment in front-end loading (try to anticipate everything that will happen or that you will need) and more investment in general knowledge, a large skill repertoire, the ability to do a quick study, trust in intuitions, and sophistication in cutting losses.

Like improvisation, a just-in-time strategy glosses, interprets, and enlarges some current event, gives it meaning, treats it as if it were sensible, and brings it to a conclusion. This form of activity looks very much like creating a stable small win (Weick 1984). And once an assortment of small wins is available, then these can be gathered together retrospectively and packaged as any one of several different directions, strategies, or policies.

Strategies are less accurately portrayed as episodes where people convene at one time to make a decision and more accurately portrayed as small steps (writing a memo, answering an inquiry) that gradually fore-close alternative courses of action and limit what is possible. The strategy is made without anyone realizing it. The crucial activities for strategy-making are not separate episodes of analysis. Instead they are actions, the controlled execution of which consolidate fragments of policy that are lying around, give them direction, and close off other possible arrange-ments. The strategy-making *is* the memo-writing, *is* the answering, *is* the editing of drafts. These actions are not precursors to strategy; they *are* the strategy.

Strategies that are tied more closely to action are more likely to contain improvisations (Weiss 1980: 401):

> Many moves are improvisations. Faced with an event that calls for response, officials use their experience, judgment, and intuition to fashion the response for the issue at hand. That response becomes a precedent, and when similar—or not so similar—questions come up, the response is uncritically repeated. Consider the federal agency that receives a call from a local program asking how to deal with requests for enrollment in excess of the available number of slots. A staff member responds with off-the-cuff advice. Within the next few weeks, programs in three more cities call with similar questions, and staff repeat the advice. Soon what began as improvisation has hardened into policy.

Managers are said to avoid uncertainty, but one of the ironies implicit in the preceding analysis is that managers often create the very uncertainty they abhor. When they cannot presume order they hesitate, and this very hesitancy often creates events that are disordered and unfocused. This disorder confirms the initial doubts concerning order. What often is missed is that the failure to act, rather than the nature of the external world itself, creates the lack of order. When people act, they absorb uncertainty, they rearrange things, and they impose contingencies that might not have been there before. The presence of these contingencies is what is treated as evidence that the situation is orderly and certain.

Conclusion

The thread that runs through this chapter is that execution *is* analysis and implementation *is* formulation. The argument is an attempt to com-bine elements from a linear and adaptive view of strategy, with a largely interpretive view (Chaffee 1985: 95). Any old explanation, map, or plan is often sufficient because it stimulates focused, intense action that both creates meaning and stabilizes an environment so that conventional analysis now becomes more relevant. Organizational culture becomes influential in this scenario because it affects what people expect will be orderly. These expectations, in turn, often become self-fulfilling. Thus the adequacy of any explanation is determined in part by the intensity and structure it adds to potentially self-validating actions. More forceful-ness leads to more validation. Accuracy becomes secondary to intensity. Because situations can support a variety of meanings, their actual content and meaning is dependent on the degree to which they are arranged into sensible, coherent configurations. More forcefulness imposes more coher-ence. Thus those explanations that induce greater forcefulness often become more valid, not because they are more accurate but because they have a higher potential for self-validation.

Applied to managerial activity, substitutes for strategy are more likely among executives because their actions are capable of a considerable range of intensity, the situations they deal with are loosely connected and capable of considerable rearrangement, and the underlying explana-tions that managers invoke (such as, "This is a cola war") have great potential to intensify whatever action is underway. All of these factors combine to produce self-validating situations in which managers are sure their diagnoses are correct. What they underestimate is the extent to which their own actions have implanted the correctness they discover.

What managers fail to see is that solid facts are an ongoing accomplish-ment sustained as much by intense action as by accurate diagnosis. If managers reduce the intensity of their own action or if another actor directs a more intense action at the malleable elements, the meaning of the situation will change. What managers seldom realize is that their inaction is as much responsible for the disappearance of facts as their action was for the appearance of those facts.

Gene Webb often quotes Edwin Boring's epigram: "Enthusiasm is the friend of action, the enemy of wisdom." Given the preceding arguments we can see reasons to question that statement. Enthusiasm can produce wisdom because action creates experience and meaning. Furthermore, enthusiasm can actually create wisdom when prophecies become self-fulfilling and factual.

One final example of a vague plan that leads to success when people respond to it and pay close attention to their response involves a religious ritual used by the Naskapi Indians in Labrador. Every day they ask the question, "Where should we hunt today?" That question is no different from, "Where is the base camp?" or "What should we do with these uniforms?" or "Should we open today?" or "Could this conceivably be the Silicon prairie?"

The Naskapi use an unusual procedure to learn where they should hunt. They take the shoulder bone of a caribou, hold it over a fire until the bone begins to crack, and then they hunt wherever the cracks point. The surprising thing is that this procedure works. The Naskapi almost always find game, which is rare among hunting bands.

Although there are several reasons why this procedure works, one is of special interest to us: The Naskapi spend most of each day actually hunting. Once the cracks appear, they go where the cracks point. What they do not do is sit around the campfire debating where the game are today based on where they were yesterday. If the Naskapi fail to find any game, which is rare, they have no one in the group to blame for the outcome. Instead, they simply say that the gods must be testing their faith.

The cracks in the bone get the Naskapi moving, just as the mountain paths drawn on the map get the soldiers moving, and just as high-tech backwardness gets Texans moving. In each case, movement multiplies the data available from which meaning can be constructed.

Because strategy is often a retrospective summary that lags behind action, and because the apparent coherence and rationality of strategy are often inflated by hindsight bias, strategic conclusions can be misleading summaries of what we can do right now and what we need to do in the future.

I do not suggest doing away with strategic plans altogether, but people can take a scarce resource, time, and allocate it between the activities of planning and acting. The combination of staffs looking for work, high-powered analytic MBAs, unused computer capability, the myth of quantitative superiority, and public pressure to account for everything in rational terms tempts managers to spend a great deal of time at their terminals doing analysis and a great deal less time anyplace else (Weick in press). It seems astonishing that one of the hottest managerial precepts to come along in some time (MBWA, management by walking around) simply urges managers to pull the plug on the terminal, go for a walk, and act like champions. One reason those recommendations receive such a sympathetic reception is that they legitimize key aspects of sensemaking that got lost when we thought we could plan meanings into existence. As we lost sight of the importance of action in sensemaking, we saw situations become senseless because the wrong tools were directed at them.

Strategic planning is today's pretext under which people act and generate meanings and so is the idea of organizational culture. Each one is beneficial as long as it encourages action. It is the action that is responsible for meaning, even though planning and symbols mistakenly get the credit. The moment that either pretext begins to stifle action meaning will suffer, and these two concepts will be replaced by some newer management tool that will work, not for the reasons claimed but because it restores the fundamental sensemaking process of motion and meaning.

REFERENCES

Bresser, R.K., and R.C. Bishop. 1983. "Dysfunctional Effects of Formal Planning: Two Theoretical Explanations." *Academy of Management Review* 8: 588–99.

Burgelman, R.A. 1983. "A Model of the Interaction of Strategic Behavior, Corporate Context, and the Concept of Strategy." *Academy of Management Review* 8: 61–70.

Chaffee, E.E. 1985. "Three Models of Strategy." *Academy of Management Review* 10: 89–98.

Chandler, A.D. 1962. *Strategy and Structure.* Cambridge, Mass.: MIT Press.

de Bono, E. 1984. *Tactics: The Art and Science of Success.* Boston: Little, Brown.

Eden, D. 1984. "Self-Fulfilling Prophecy as a Management Tool: Harnessing Pygmalion." *Academy of Management Review* 9: 64–73.

James, W. 1956. "Is Life Worth Living?" In *The Will to Believe,* edited by W. James, pp. 32–62. New York: Dover.

Kerr, S., and J.M. Jermier. 1978. "Substitutes for Leadership: Their Meaning and Measurement." *Organizational Behavior and Human Performance* 22: 375–403.

McCaffrey, D.P. 1982. "Corporate Resources and Regulatory Pressures: Toward Explaining a Discrepancy." *Administrative Science Quarterly* 27: 398–419.

McCall, M.W., Jr., and R.W. Kaplan. 1985. *Whatever It Takes: Decision Makers at Work.* Englewood Cliffs, N.J.: Prentice Hall.

Snyder, M. 1984. "When Belief Creates Reality." In *Advances in Experimental Social Psychology, Vol. 18,* edited by L. Berkowitz, pp. 247–305. New York: Academic Press.

Snyder, M., E.D. Tanke, and E. Berscheid. 1977. "Social Perception and Interpersonal Behavior: On the Self-Fulfilling Nature of Social Stereotypes." *Journal of Personality and Social Psychology* 35: 656–66.

Staw, B.M. 1980. "Rationality and Justification in Organizational Life." In *Research in Organizational Behavior, Vol. 2,* edited by B.M. Staw and L.L. Cummings, pp. 45–80. Greenwich, Conn.: JAI.

Thompson, J.D. 1964. "Decision-making, the Firm, and the Market." In *New Perspectives in Organization Research,* edited by W.W. Cooper, H.J. Leavitt, and M.W. Sheely II, pp. 334–48. New York: Wiley.

Warwick, Donald P. 1975. *A Theory of Public Bureaucracy: Politics, Personality, and Organization in the State Department.* Cambridge, Mass.: Harvard University Press.

Watzlawick, P., J.H. Beavin, and D.D. Jackson. 1967. *Pragmatics of Human Communication.* New York: Norton.

Weick, K.E. 1979. *The Social Psychology of Organizing,* 2d ed. Reading, Mass.: Addison-Wesley.

———. 1984. "Small Wins: Redefining the Scale of Social Problems." *American Psychologist* 39: 40–49.

———. 1985. "Cosmos vs. Chaos; Sense and Nonsense in Electronic Contexts." *Organizational Dynamics* 14 (Autumn): 50–64.

Weick, K.E., D.P. Gilfillan, and T. Keith. 1973. "The Effect of Composer Credibility on Orchestra Performance." *Sociometry* 36: 435–62.

Weiss, C.H. 1980. "Knowledge Creep and Decision Accretion." *Knowledge: Creation, Diffusion, Utilization* 1(3): 381–404.

11 CONCLUSIONS

David J. Teece

There is little doubt that the chapters in Part I of this volume contain important guidance for managers and analysts. For instance, Porter's distinction between global and multidomestic industries is very helpful. Disaggregating economic activity according to the value chain brings into sharp relief both location and integration issues. The role of cooperative agreements—as well as hazards—is brought into sharp relief. What Porter offers is a set of tools for evaluating international competition and for organizing a competitive response. Clark offers a framework for assessing how innovation will affect a firm's existing capabilities. The weaknesses of existing established techniques for project evaluation are exposed; and a framework is offered for assessing the relative impact of a new technology on industry participants.

Wheelwright puts U.S. declining manufacturing competitiveness in stark focus—it is primarily a problem of management and organization, not a problem of high wages. Serious flaws in existing practices are exposed. A new approach is advocated that can make U.S. manufacturing competitive. Lawrence points out that adjustment is not easy, particularly when there is excessive competition—a concept derived not from economics but from the social psychology of organizations. Two of these chapters have some common themes: U.S. competitive performance has been impaired by management practices that have not been as attuned as they might be to the challenges of international competition. Increased competition is exposing these weaknesses, but not enough is being done to correct them. A refocusing of attention, prestige, and incentive to manufacturing and innovation is in order in both our companies and our business schools. Labor must also begin to play a more positive role in productivity improvement.

Part II contains several strikingly original contributions to the literature on strategy and organization. Pfeffer issues a strong challenge to much of the strategy literature, pleading for scholars to bring the environment back into the analysis and to recognize, in Winter's term, that many (business) environmental factors are control and not state variables. The business environment to which Pfeffer refers is increasingly international in scope; competitive success will therefore increasingly depend on firms' ability to manage their external relations abroad as well as at home. The lack of appreciation on the part of U.S. management for some of the considerations raised by Pfeffer is revealed by the fact that the only U.S. manufacturing industry to have a trade association office in Tokyo is the semiconductor industry!

Rumelt's effort to outline a theory of entrepreneurship is noteworthy both for its ambition and its success. In a strikingly original treatment, Rumelt shows a distinction between entrepreneurial rent and conventional forms of rent. Entrepreneurial rent is traced to *ex ante* uncertainty and discovery; it is protected by causal ambiguity as to its foundations, as well as first-mover advantages and other factors. Important implications for public policy, in particular antitrust policy, are cited.

Winter pursues some of the same issues Rumelt is concerned with, and similarly introduces several new concepts into the strategy literature. From control theory comes the distinction between state and control variables and the concept of full imputation. The concept of the heuristic frame is used to explain some of the fads and fashions in the strategy literature.

In a concluding chapter, Weick argues that implementation is what strategy is all about. Too much strategy can paralyze an organization. Presumably too little can have the same effect if the firm is unable to perceive threats until they occur, or if it lacks the ability to think through major strategic choices. Strategic planning cannot, however, substitute for action because action is necessary not only to get things done but also to experiment, learn, and create meaning.

There is little doubt that Weick is fundamentally correct; the best consultants and managers recognize the importance of action and implementation. It is all in the details and how they are performed. But it is hard to believe that forays into planning and analysis are necessarily mischievous. There may be substitutes for strategy, but the real thing is often effective, particularly when the costs of experimentation (such as committing billions of dollars to a development program) are high.

This volume makes no pretensions with respect to completeness. Important issues in finance, marketing, and operations have been swept aside to explore normative as well as theoretical issues in strategy and organization. Several of the contributors are not mainstream scholars in strategy; their positive contributions illustrate the mileage to be gained from bringing external disciplines into the strategy literature. It is hoped that the efforts in these pages will make a modest contribution to our understanding of the development and performance of the business enterprise and ultimately to the competitiveness of the U.S. economy.

ABOUT THE CONTRIBUTORS

Kim Clark
Professor of Business Administration
Harvard University

Paul R. Lawrence
Wallace Brett Donham Professor of Organizational Behavior
Harvard University

Jeffrey Pfeffer
Thomas D. Dee II Professor of Organization Behavior (and by courtesy in
    the Department of Sociology)
Stanford University

Michael E. Porter
Professor of Business Administration
Harvard University

Richard P. Rumelt
Associate Professor of Competitive Strategy
University of California at Los Angeles

David J. Teece
Professor of Business Administration and Director, Center for Research
    in Management
University of California at Berkeley

Karl E. Weick
Professor, Harkin & Co.
  Centennial Chair in Business Administration
University of Texas

Steven C. Wheelwright
Kleiner, Perkins, Caufield, and Byers Professor of Management
Stanford University

Sidney G. Winter
Professor of Economics and Professor, School of Organization and
  Management
Yale University

NAME INDEX

SUBJECT INDEX